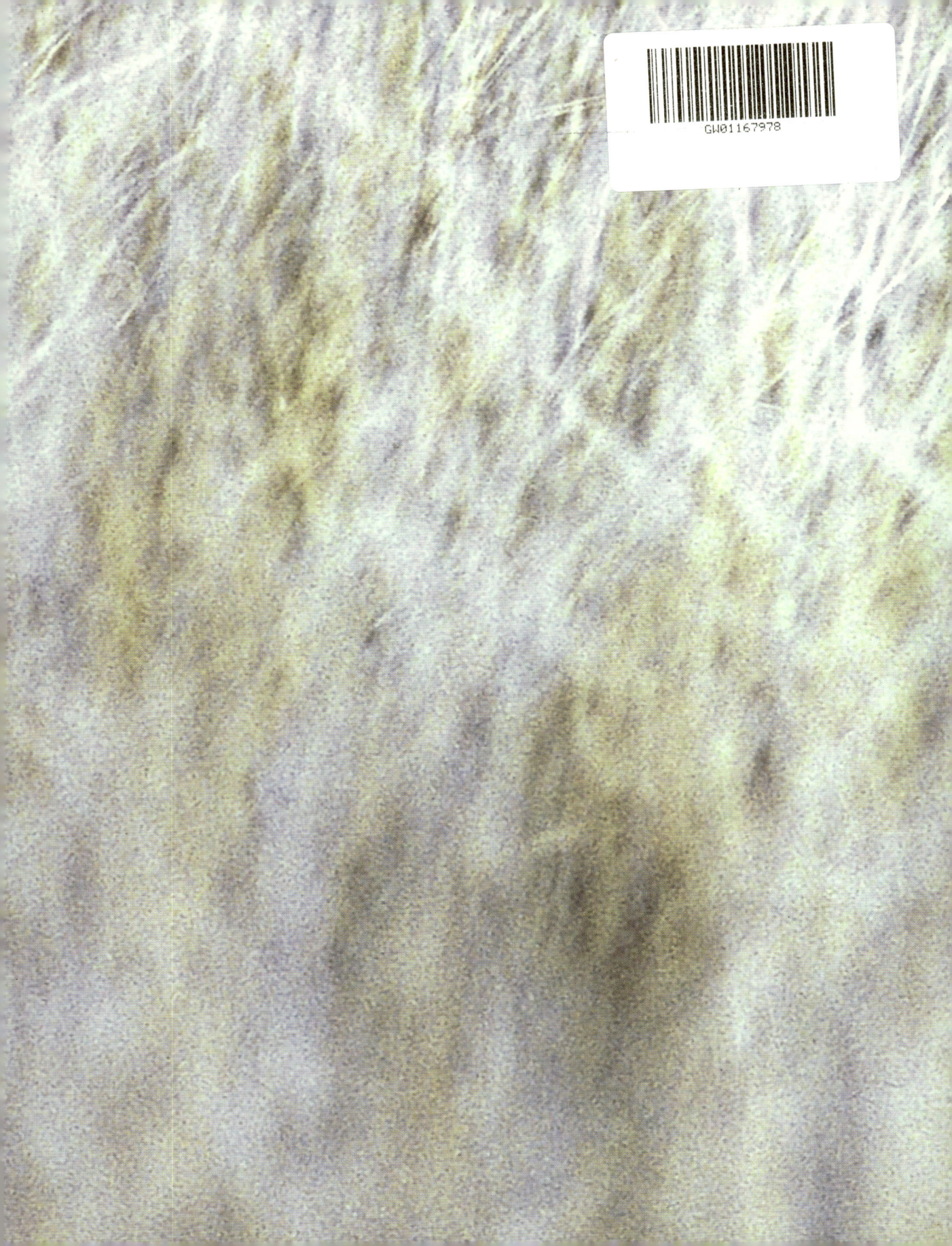

Ten Thousand Acres

A LOVE STORY

Ten Thousand Acres

A LOVE STORY

PATRICE NEWELL

Photography by
Simon Griffiths and
Travis Peake

LANTERN
an imprint of
PENGUIN BOOKS

LANTERN

Published by the Penguin Group
Penguin Group (Australia)
250 Camberwell Road, Camberwell, Victoria 3124, Australia
(a division of Pearson Australia Group Pty Ltd)
Penguin Group (USA) Inc.
375 Hudson Street, New York, New York 10014, USA
Penguin Group (Canada)
90 Eglinton Avenue East, Suite 700, Toronto ON M4P 2Y3, Canada
(a division of Pearson Penguin Canada Inc.)
Penguin Books Ltd
80 Strand, London WC2R 0RL, England
Penguin Ireland
25 St Stephen's Green, Dublin 2, Ireland
(a division of Penguin Books Ltd)
Penguin Books India Pvt Ltd
11 Community Centre, Panchsheel Park, New Delhi – 110 017, India
Penguin Group (NZ)
Cnr Airborne and Rosedale Roads, Albany, Auckland, New Zealand
(a division of Pearson New Zealand Ltd)
Penguin Books (South Africa) (Pty) Ltd
24 Sturdee Avenue, Rosebank, Johannesburg 2196, South Africa

Penguin Books Ltd, Registered Offices: 80 Strand, London, WC2R 0RL, England

First published by Penguin Group (Australia), a division of Pearson Australia Group Pty Ltd, 2006

10 9 8 7 6 5 4 3 2 1

Text copyright © Patrice Newell 2006
For copyright in the photographs, refer acknowledgements page

The moral right of the author has been asserted

All rights reserved. Without limiting the rights under copyright reserved above, no part of this publication may be reproduced, stored in or introduced into a retrieval system, or transmitted, in any form or by any means (electronic, mechanical, photocopying, recording or otherwise), without the prior written permission of both the copyright owner and the above publisher of this book.

Design by Sandy Cull © Penguin Group (Australia)
Cover photograph by Travis Peake
Typeset in Mrs Eaves by Post Pre-press Group, Brisbane, Queensland
Printed and bound in China by 1010 Printing International Limited

National Library of Australia
Cataloguing-in-Publication data:

Newell, Patrice, 1956– .
Ten thousand acres: a love story.

ISBN 1 920989 16 1.

1. Land use, Rural – New South Wales – Upper Hunter Region. 2. Food crops – New South Wales – Upper Hunter Region. 3. Human geography. 4. Upper Hunter Region (NSW) – Rural conditions. I. Title.

630.99442

www.penguin.com.au

For if the entire history of landscape in the West is indeed just a mindless race toward a machine-driven universe, uncomplicated by myth, metaphor, and allegory, where measurement, not memory, is the absolute arbiter of value, where our ingenuity is our tragedy, then we are indeed trapped in the engine of our self-destruction.

Simon Schama, *Landscape and Memory*

We humans were originally hunters and gatherers. Next we became herdsmen. Then we began to plant and harvest crops. It was farming that provided the foundation for civilisations: for cities, pyramids, cathedrals.

This process, which lasted thousands of years elsewhere, has taken less than two centuries in Australia. In only a few generations we have gone from an agricultural age to a knowledge age, but this knowledge is more concerned with cyberspace than with soil, and it has little respect for the land.

Here at Elmswood we've spent the past twenty years trying to slow down, just a little, the relentless pace of change, and to heal some of the wounds inflicted by farming practices in the past. This is the story of those years.

Our farm, Elmswood, arcs around Gundy in the Upper Hunter Valley of New South Wales. The little town was once so photogenic that it was used as the location for feature films. Our 1890s homestead sits at the junction of two rivers, the Isis and the Pages, on a low knoll overlooking flats of rich alluvial soil. To the west, hills with ironbark and box trees rise gently; beyond them, massive geological thrusts have piled the landscape into wilderness, peaking with Black Mountain, sometimes dusted with snow.

Beef cattle graze most of the land, sharing it with kangaroos, wallabies, wallaroos, echidnas, goannas, snakes, lizards, turtles, and many more creatures. We also keep about sixty ewes to supply us with fat lambs to eat. A hundred hectares have been planted with olive trees, from which we produce oil and soap. Below the homestead we've irrigated ten hectares to grow lucerne for hay and silage. We gather eggs, jar honey, harvest vegetables.

Sometimes I look at the trees in the olive grove and imagine what they'll be like in my old age – gnarled but still productive. We will grow old together. My commitment to this land is far stronger than when we first took ownership. Then my feeling for it was wondering, awed, tentative. Now, like my olive trees, it is deeply rooted, convinced. Every day we are conscious of living in beauty, and the mysteries of the place continue to reveal themselves. It's as if the farm is a huge scroll that we continue to unroll, disclosing more of the painter's vision, both in the panoramic and the particular.

With my partner Phillip and our daughter Aurora

5
Ten Thousand Acres
A Love Story

Today much of this ancient land that evolved so quickly into an agricultural and pastoral landscape after white settlement is being lost to development, and to ecological disasters. Family farms have been hacked into fragments or merged into corporate enterprises. With this has come a loss of continuity and community. More importantly, it spells catastrophe for native biological diversity.

Ever fewer people are engaged in food production, and farming is becoming an increasingly distant and old-fashioned idea. Many people think of Australian agriculture as an immense mistake. They know the bitter consequences of ignorance, of disregarding the land's limitations. It's true that some farmers have been blind and arrogant — in economic terms, destroying our capital; making our soil hard, naked, saline. Turning earth into sand.

Little wonder, therefore, that agriculture is shown scant respect. Less and less beautiful, decreasingly natural, farms are being forced by the demands of the mass market to become more and more like factories. Farmers, once proud, now feel like serfs to the czars of the shopping mall.

Yet not all agriculture is a lost cause. And not all farmers are the land's enemies. Many of us are rethinking basic assumptions, and while we might be outnumbered we're determined to turn the tide. We farm with new ambitions, never forgetting that there is no degree of separation between us and nature. We are its children. And every portion of our farm, every tiny piece of it, is part of the whole.

Some areas may not be used for economic gain but they remain important to the ecosystem — not only to the rhythm of the seasons but to the great cycles of water, minerals and energy. To be a good farmer you must keep the land alive.

Once, others lived with this landscape, with all the landscapes within it. They had a kinship earned over millennia. They knew the whispers in the wind, the hints dropped by leaves, the urgent messages in the behaviour of birds, animals, even insects. They spoke the language of landscape. While their understanding was not scientific, it had absolute precision.

Those people have been driven off, their knowledge now inaccessible to me, leaving questions that I try to answer with the reference books I carry in the valleys and hills. We discover next to nothing. At best we rediscover.

As with friendship, learning this land needs time. Here a muster, a harvest, a new fence, a tree planted, a picnic, a swim, a climb up a tree. There the picking of a flower for pressing, a search for missing cattle, the taking of a photograph or a wrong turn. I won't let these places be unremembered.

Elmswood homestead, circa 1906

One autumn we climb into the foothills to see what really grows on a south-facing slope. It won't be all work, it will be fun too. Travis, a friend and a trained botanist, his botanist partner Liza and their two-year-old son Nicholas will be coming.

We set up camp beside Kewell Creek — a rivulet of crevices, secrets and shrouding she-oaks — in a place intercepted by a dramatic gully where rock pools are deep and the older trees have mostly toppled over. In falling, many have reached out to their relatives on the other side of the creek, forming bridges that beg to be crossed.

It is not quiet here. The valley is full of birdsong and we enjoy eavesdropping. Voices on the wing. Birds soaring above high C. Our spirits soar with them. Parrots whip past making a breeze amid the complaints of those critics the currawongs.

While Aurora chases Nicholas like a mother hen her chick, we unload the truck, arrange stones in a circle for a campfire, set up the card table and unpack cameras, books, food. We spread our rugs over grass as spiky as echidna quills and open the picnic basket, but Aurora leaves the rug in favour of a wildflower patch and lies on her back looking at the sky, the gloss of her hair the colour of last year's grass.

The rocks in the creek, small, middle-sized and immense, get shifted in every flood. Over the centuries, they have crowded together, forcing the water to go around, under and over them, creating a music of movement that rises to meet the she-oak murmur.

We're in Ridge Paddock, so named because of a mighty spine that divides two deep, dramatic flanks. Up until two months ago, I had cattle here to keep the grass and weeds down. In their absence the paddock has become a horticultural display, with yellow *Ranunculus lappaceus* along the verges of the gully. Their waxy petals, bright and showy, dangle from thin stems. Further up the hillside, clinging to rocky outcrops, the riotous purple of *Verbena litoralis* challenges the yellow.

I find a legume-looking leaf and ask Travis if it's a native. It is, and he tells me that farmers should be planting them more, instead of exotics. Australia has two thousand native legumes, yet most farmers know nothing of them, concerning themselves in the main with a few foreigners, such as lucerne and Haifa white clover. There's a native legume called *Trigonella suavissima,* Cooper's clover, that grows mainly inland, but perhaps we can find some around here.

Close by, having colonised a patch of black basalt turned over by feral pigs, onion weed appears in clumps. Nicholas has tumbled onto them and the aroma of onion wafts out. Breaking off a piece, he holds it up to Liza. When he stands the plant bounces back upright, aquiver.

Time to boil the billy. The fallen logs littering Ridge Paddock are damp from recent rains and Aurora leads Nicholas off in search of drier wood. While they pile the kindling, I begin collecting: an interesting bone, a bird's nest, a white pebble that might be marble, a rock so spherical it could be a cannonball.

We organise ourselves to cut plant samples. I'm learning to be systematic, to position cuttings neatly in old newspapers, with clear notification of where and when they were taken and with an amateur's guess at identification. Back home, though, I'll be forensic in my accuracy.

Where the steepness of the land has discouraged stock, the weed count diminishes. Travis pegs out an area ten by ten metres, hammering white builder's pegs into the ground. I activate my GPS and note their position. We pull yellow tape around the pegs, I date my notebook and we begin to count the species. It's the same technique used by archaeologists, or police investigating a crime scene. We know from past experience how easy it is to overlook the evidence, to miss plants right before our eyes.

'You can tell a lot about a person when you do this,' Travis says. 'About their stamina, their ability to focus, how easily they tire and move on, whether they have an eye for distinguishing.' He tells stories about botanists who go to a site, guess its composition and try to prove it.

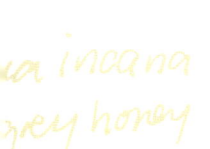

Their premise is that if it looks depleted it probably is.

Aurora helps while Nicholas plays in a puddle. The soft low light in these late hours of sunshine stretches into a slow sunset. No frown-inducing brightness to turn you away. Protected by the shadow of the hills, flowers stand erect, as if proud of our attention.

Some 2 per cent of the ground in our little patch is bare and I calculate the litter at 8 per cent. There is good grass coverage, the slope is moist, the soil is a red clay, run-off will be slow, the area has been recently grazed.

We're surprised to find forty-three different plants in such a small space. They're humble efforts, not the sort that would have a bright future in a commercial nursery. Nor would they have thrilled those plant hunters of the eighteenth and nineteenth centuries who were intent on claiming prizes for the likes of Kew Gardens. In those days botanists travelled the world, and all the empires — the English, French, Portuguese, Spanish, Dutch — had their own trophy-seeking teams. Some went in search of beauty that could be traded, others were looking for medicines. The accounts of their journeys were enthusiastically awaited in their lifetimes and are being republished today. But they would not have waxed lyrical about our forty-three species.

Travis Peake

We, however, do. We rate the quantity of each with a coding from one to six — a cover-abundance, or CA, score — with one being rare (accounting for less than 5 per cent of an area) and six being plentiful (between 75 and 100 per cent).

Later, we repeat this exercise on other parts of the farm, and grow to treat these little ventures as epic journeys. They don't take us far from home — perhaps fifteen kilometres at the most, an hour-and-a-half from our back door. But after every trip, these hidden worlds feel further and further away, as if we've been travelling in foreign countries.

Chloris truncata
windmill grass

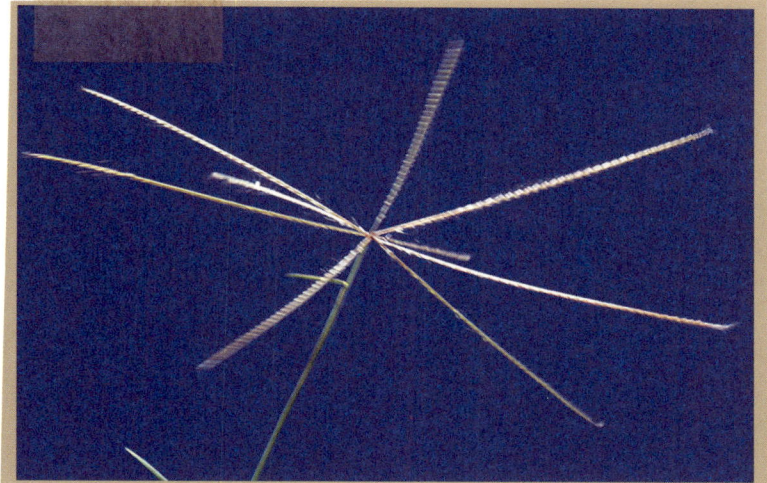

Brachyscome sp.

Carthamus lanatus
saffron thistle

Verbena rigida var. *rigida*
veined verbena

13
Ten Thousand Acres
A Love Story

The Plants at Kewell Creek

Alternanthera pungens, khaki weed (1)
Anagallis arvensis, scarlet pimpernel (1)
Aristida ramosa, wiregrass (3)
Austrostipa scabra, corkscrew grass (2)
Bothriochloa macra, red grass (3)
Brachyscome multifida, cut-leaved daisy (2)
Calotis lappulacea, yellow burr-daisy (2)
**Carthamus lanatus*, saffron thistle (2)
Cheilanthes distans, bristly cloak fern (1)
Cheilanthes sieberi subsp. *sieberi*,
   poison rock fern (2)
Chloris truncata, windmill grass (2)
Chloris ventricosa, tall chloris (2)
**Conyza sumatrensis*, tall fleabane (2)
Cymbopogon refractus, barbed wire grass (1)
Cynodon dactylon, couch (1)
Cyperus gracilis, sedge (2)
Daucus glochidiatus, native carrot (2)
Desmodium varians, slender tick-trefoil (2)
Dichanthium sericeum subsp. *sericeum*,
   Queensland bluegrass (4)
Dichondra repens, kidney weed (2)
Einadia trigonos, fishweed (1)

Elymus scaber var. *scaber*, common wheatgrass (1)
Eragrostis brownii, Brown's lovegrass (2)
Eriochloa pseudoacrotricha, early spring grass (2)
Geranium solanderi var. *solanderi*,
   native geranium (2)
Glycine tabacina, glycine (1)
**Lolium perenne*, perennial ryegrass (1)
Mentha satureioides, creeping mint (2)
Minuria leptophylla, minnie daisy (1)
Oxalis exilis, wood sorrel (2)
Panicum effusum, hairy panic (2)
Pelargonium inodorum, wild geranium (1)
**Petrorhagia velutina*, velvet pink (2)
Phyllanthus virgatus, spurge (2)
Pimelea linifolia, slender rice flower (1)
Portulaca oleracea, pigweed (2)
Pseuderanthemum variabile, pastel flower (2)
Rumex brownii, swamp dock (2)
**Senecio madagascariensis*, fireweed (1)
Sida filiformis (2)
Sporobolus creber, slender rat's tail grass (3)
**Verbena litoralis*, coastal verbena (1)
Wahlenbergia sp., native bluebell (2)

The numbers in brackets show the cover-abundance score. An asterisk denotes a plant not native to Elmswood. A complete flora list for Elmswood can be found in the appendix.

Windmill Grass
Airstrip Paddock.
Chloris truncata
Summer 2002.

The early colonists wanted grasses from home to plant. Familiarity bred contentment. Others argued against such sentimentality, urging caution. Let's see what the native grasses can do, they suggested. Could they nourish the imported animals? Or be used for crops?

If only those grasses could have been our wheat or rice. Now, belatedly, science is assessing them. But it will take a lot to persuade Australians to care as much about a paddock of native grasses as they do about a vista of wheat or corn.

The exotic grains sprout and thrust for harvest; they're as speedy as racehorses. They're also fully understood, with thousands of years of growing and knowing behind them. Some days our native pastures look like costumes we're trying on. As we learn to appreciate them, perhaps one day we'll be able to say, 'Doesn't that field of wallaby grass look brilliant?' Or, 'See how the hairy panic dances at sunset.' Or, 'Hear how the Queensland bluegrass sighs.' And will I be able to make kangaroo flour from kangaroo grass to bake kangaroo bread?

These days, native grasses are only on the menu for cattle, but if our grains could be harvested from our perennial species the land would be the better for it. I take heart from the fact that more than a thousand people are members of native grassland associations across the country.

Themeda australis
kangaroo grass

My mother, Thelma, was seventy-two when she moved to the farm with us. She remembered riding in a jeep in Central Australia when she was in her twenties, and spoke of it as the biggest adventure of her life.

Our first four-wheel drive was a vomitous yellow Nissan. With Mum in the back seat nursing Willie, her Jack Russell pup, we took the truck on its inaugural drive to a high point of the farm, eight hundred metres up Black Mountain. Along overgrown tracks, past hillsides covered in rubble and across gaunt ridges we bumped, seatbelts jamming, Willie barking rapturously, Mum strangely silent until we got to the top with its 360-degree view. Whereupon she threw herself onto the ground and, like His Holiness at the airport, kissed it. Really kissed it.

This became our first sacred place. The place where Mum always wanted to take her friends so she could tell them of her terror and say, 'This is where I kissed the earth.'

Thirteen years later, Mum died quietly at home. We took the urn with her ashes up the mountain, driving in silence to her spot, and gathered rocks to form a circle around the place she'd kissed and scattered her ashes there.

Ever since, on every trip to the top, Aurora, Phillip and I stop to add rocks to what has slowly become a cairn. Mum would have enjoyed its slow creation. My mother never rushed. We sit together sharing the view, sharing our memories of Thelma.

We don't do this on trips to the top with unsympathetic souls who might think our ritual amusing, but it has become a profound and loved tradition.

Once in a while it snows on our mountain. Not very much, not very deeply, and not the panoramic blue crystalline snow of the ski fields. Rather a puff of lightness, a talcum-powdering of the trees. Invariably this is a time of great excitement. We put on beanies and scarves and slither up the steep tracks on quad bikes. At the peak we look out on a calm white coldness. We hear its colour.

After one snowfall, bright berries lay scattered over the snow, leaving vivid bloodstains. From our rocky outcrop the snow-capped peaks on the far horizon made dry and dusty Australia look like Canada.

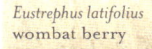

Eustrephus latifolius
wombat berry

Our new table is a masterpiece. We acquired it from Barry and Stephen, a father and son who specialise in recycling old buggies and farm equipment, turning anachronisms into furniture and sculpture. In nearby Scone they discovered a derelict dairy shed, the massive planks of which had been painted and repainted over a century, concluding with an improbable pistachio green. Barry carefully released the planks and scrubbed them down to make tables of various dimensions, and instead of sanding off the pistachio, he beeswaxed the wood until it gleamed. A green gleam.

Phillip, stopping at the shop to fossick through their rural memorabilia, returned excited by his find. God knows, we don't need another table but we want the sheen of the planks and the feeling of history.

I cover its silken surface with my paraphernalia, carefully laying out recently collected plant samples. And press ahead with the pressing. In summer, leaves and flowers dry rapidly but today, with the air moist from a light shower, the plant juices seep into the newspaper, slowing their transformation into everlasting specimens.

23
Ten Thousand Acres
A Love Story

I take time to reposition a flower head, a petal; to open out a leaf. It's as if I'm laying out a corpse. The final position. I fold a stem of slender rat's tail in a zigzag so that it will fit neatly across the sample sheet.

My mother used to spend her afternoons strolling to the shearing shed with a dog or two, then up past the cattle yards to collect plants to bring home and arrange. She'd put them in an old, unpolished brass vase — Aladdin-like, with Moorish curves — that she'd perch on top of the television.

When there weren't any flowers to pick she'd turn to grasses. Dry already, they made spectacular, enduring arrangements. Barbed wire grass, wallaby grass, wiregrass, Queensland bluegrass, slender rat's tail, and the wonderfully named hairy panic, its honey-coloured stems as fine as hair.

When Mum decided it was time to move from the farm into a small flat in town she took her vase of grasses with her. On her return to Elmswood for the last months of her life we brought it back. Sorting her things after her death, I lifted her arrangement from the Aladdin vase and tossed it to the winds in the olive grove. Though I did this tenderly, I've ever since felt regret. My mother was excited by grasses before I'd truly discovered them, and I abandoned her last collection.

Now I make amends by collecting grasses for her vase.

Slender rat's tail
Sporobolus creber.

- poor nutrition
  handles heavy grazing.

- round leaf sheath.
  smooth " "
  hairless " "

- silk hairs near base

8/5/04
Tank Pd

Land ownership involves responsibility. It doesn't matter if you have a residential block in the city or a million acres. The bigger the scale, the greater the responsibility — and the burden. Sometimes the problems involved in agriculture are so immense they feel as if they could crush you. Whenever they threaten to, I remind myself of John Keats' notion of 'negative capability'. He was referring to literature, not land, but he wanted the mind to simply *be*. To be in doubt and mystery without anxiously grappling for answers. For Keats, doubt led to insight into what it is to be human. He has taught me that difficulties, even sadness, can be useful and are not to be feared.

Methods for converting negativity into positivity crowd the shelves of bookstores in volumes of psychobabble and self-help. I prefer Keats to Oprah Winfrey. Instead of rejecting doubt, explore its depths.

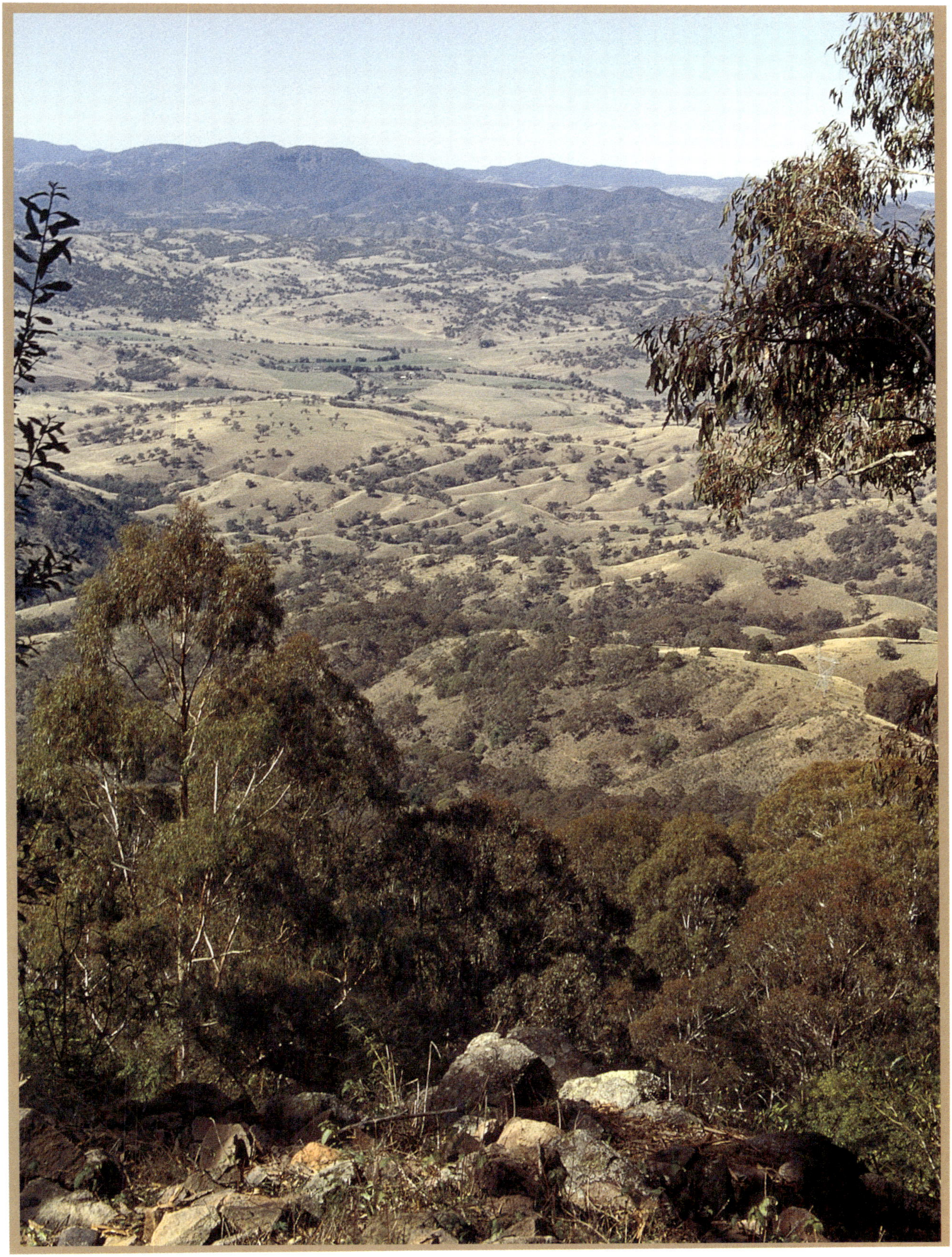

Echinopogon caespitosus
var. *caespitosus*
tufted hedgehog grass

You can feel it in the air when another big dry is approaching. The paddocks around the yards are brimming with cattle to be sorted. Before we can load them into trucks, they eat every blade of grass and the earth looks like it's taken off its clothes, revealing the brown dust we first encountered when we arrived in the summer of 1987.

Back then, much of the farm was like that. We ploughed, sowed fancy new hybrid seeds, sprayed our biodynamic preparation, and watched devastated paddocks become productive again.

After a particularly bad drought, I decided to leave these flat paddocks by the yards alone, to give them a rest. Not even a single horse grazed there. The land couldn't take any more. The heat would cause new germinations to wither.

Two years later, a miracle happened. After a shower of gentle rain, the whole forty hectares became a haven for native grasses. *Austrodanthonia fulva*, or wallaby grass; *Themeda australis*, kangaroo grass; *Dichanthium sericeum* subsp. *sericeum*, Queensland bluegrass; *Austrostipa scabra*, corkscrew grass; *Bothriochloa macra*, red grass; *Echinopogon caespitosus* var. *caespitosus*, tufted hedgehog grass.

Here, beside the most intensively worked area on the farm, the old plants returned, bringing subtle tones of green. These ancient, unfashionable grasses turned out to be as nutritious for cattle as the high-tech grasses so enthusiastically marketed by the big seed companies.

This is how it should be, I thought, inspecting the paddocks, and I decided to do this everywhere. To create vast billowing fields of native grasses.

Using filaments of electric fencing wire, we began to divide a giant paddock into smaller ones, moving stock from one cell to another. One paddock became two, four, eight, and by choreographing the movement of cattle we have protected and enriched the grasses.

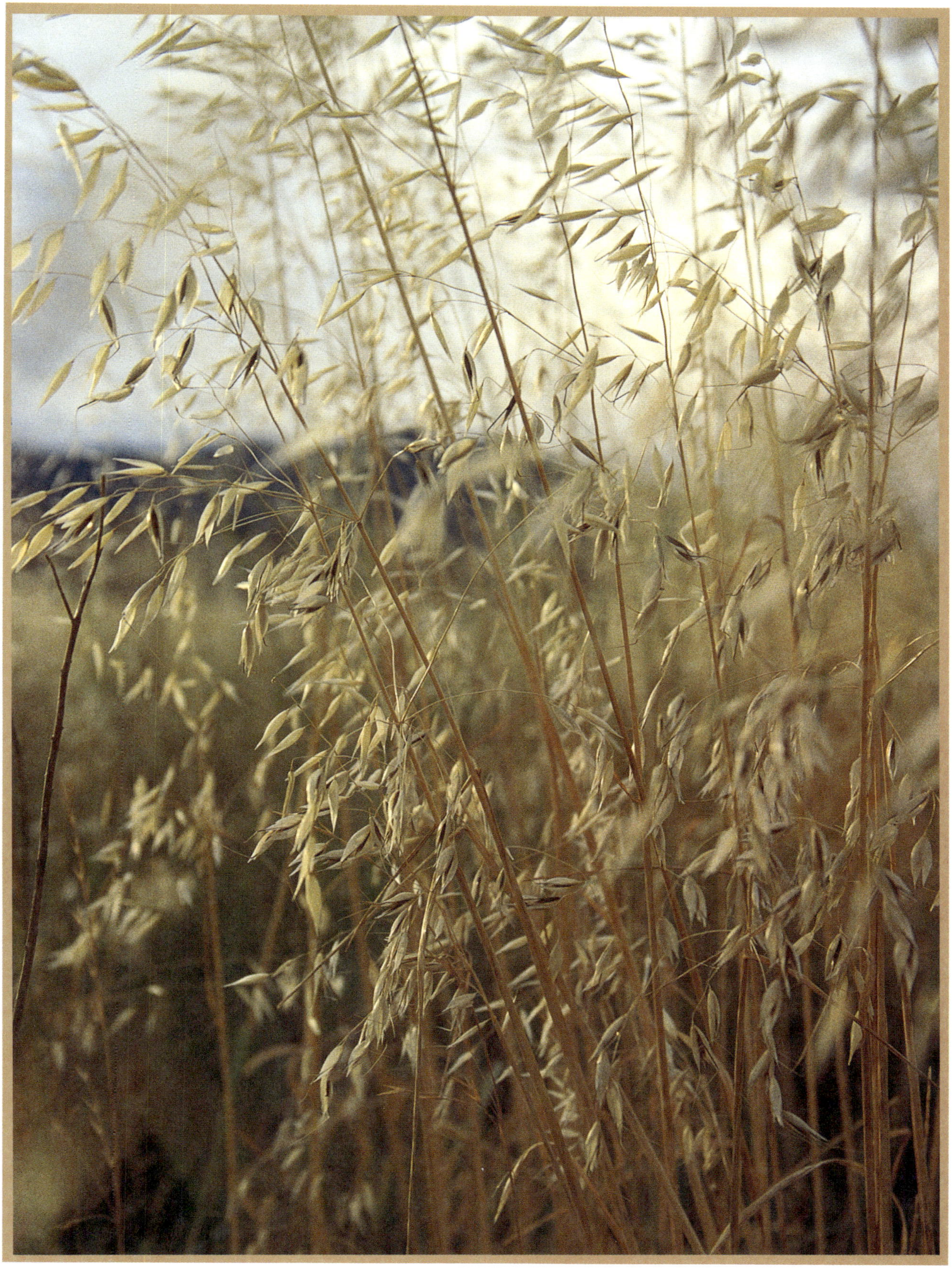

The geebung, *Persoonia linearis*, signposts the track as we head up to Black Mountain

Parts of Black Mountain look like a war zone. They're littered with skeletons, regiments of dead trees killed by past owners with poisons, axes and government permission in order to open up land for cattle and sheep. A recent storm has skittled more dead trees, and in accordance with Murphy's law, nine out of ten have fallen over a pathway or a track.

On the north-facing side of the mountain we split fallen timber, stacking it into heaps, creating space for new trees to sprout — young recruits in green uniforms. Bushfire management is difficult here. In the event of a blaze started by lightning, the whole mountain could explode and there'd be no stopping it. Taking cues from cattle tracks and roo paths, we improve access and little by little try to regenerate an area.

A hundred years of sheep-grazing and decades of spreading superphosphate have combined to remove the native grasses. We sprinkle *Austrodanthonia* seed lightly over the disturbed earth, hoping it will sprout. So far it hasn't. Even where Black Mountain has escaped the axe, where the dry sclerophyll forest remains, it is more haunting than beautiful.

Wanting to capture the madness of our forebears, Travis stops to take a photograph of ringbarked trees, his fingers tracing an axe mark. Although I wasn't responsible for this desolation, I feel ashamed. Looking straight ahead as Travis wades with his camera through the horehound and sticky-beak, without moving my head an inch, I can count a hundred murdered trees.

Can time heal everything?

32
Ten Thousand Acres
A Love Story

I used to think that meditation was another word for prayer. But prayer in my Catholic days was a time to rest and reflect. In contrast, much of the meditation practised today seems narcissistic, a way of disengaging from the important issues. It is healthy, finally, to face what is happening around you in the world, rather than focus solely on your own problems or enjoyment.

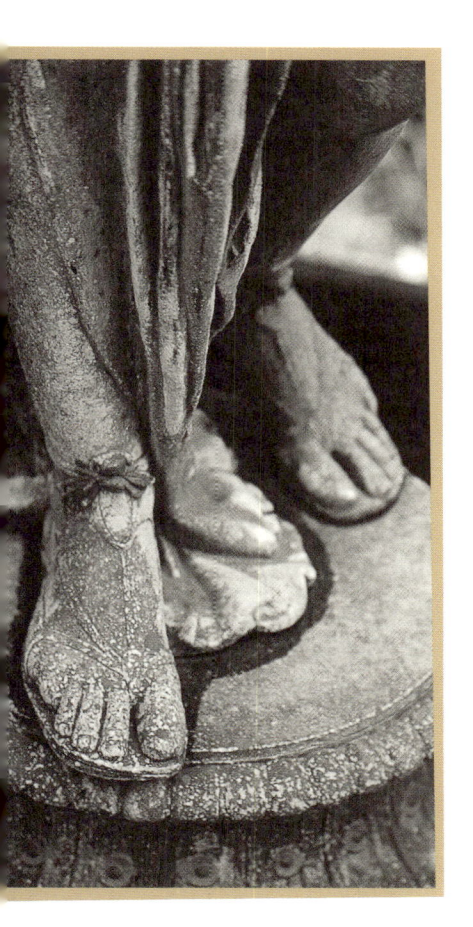

For nearly twenty years, every Thursday morning, punctually at eight, my friend and distant neighbour Yvonne has arrived to help around the house. As the years have passed, we've shared grief over the death of parents, worried about the illnesses of partners, discussed our dogs, bemoaned our aches and pains, and through all our troubles have known we share a fortunate life.

Once a month we climb the stairs to my office to concentrate on the accounts. Of all the farm jobs, this can be the most oppressive, yet Yvonne makes it almost pleasurable. She too manages a farm, and we commiserate about increased costs and the ever-mounting pile of paperwork demanded by government rules and regulations. We half suspect that much of it gets fed into a bureaucrat's shredder.

We exchange surplus vegetables and swap things the other can't grow. (I've given up on rhubarb, she doesn't bother with asparagus.) Yvonne never wastes anything, not time, not words, not burnt pots, tea boxes or jars. Old pots make good chook feeders; you can store nails in tea boxes, and reuse jars for preserves. I once discovered her stripping the wire from an old electric blanket, because, she said, it would make a good picnic rug.

We appreciate each stitch in the tapestry. When the earth is ploughed, when a calf is born, an olive ripens, a bed is made or a table cleared, we make it matter. Everything matters.

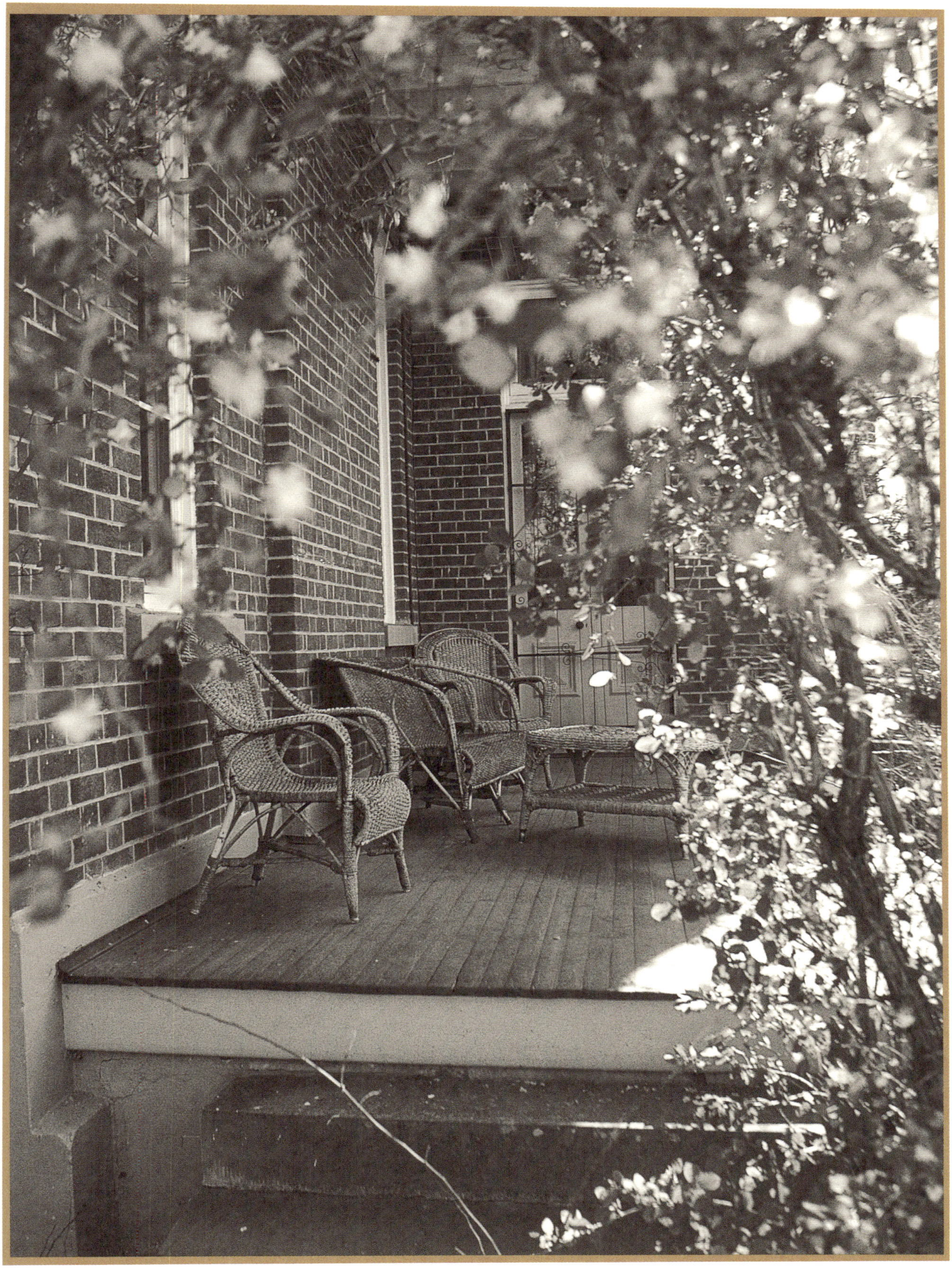

At the ending of a day the sunset emphasises the wounds of an eroded hillside. I add that to the endless list of tasks to be tackled. We don't have a lot of erosion but I'm determined to have less. The land may be voiceless but I am listening.

NOVEMBER NOTES

In city jobs, in professional life, it is hard to survive your mistakes. Bosses sack, patients sue, ratings fall, tenure is terminated. Elmswood is more forgiving of our errors. Plants and animals adapt to my shortcomings and to the harshness of the climate.

I can imagine the desolation farmers must feel when they're forced from their land because of financial problems, the appetites of greedy developers or hungry miners, or by their own failing health. Whatever compensation they might get in cash, there's no recompense for that sense of utter defeat.

I think of the farmers who preceded me. Generations of white families have lived here, each investing so much more than money. Increasingly I meet farmers reconciled to the inevitability of loss. The income isn't sufficient, the droughts are deepening, the creeks aren't flowing, the kids aren't interested, the area is being chopped up for lifestyle blocks, market share is being gobbled up by agribusinesses in unholy partnerships with food companies and supermarkets. Rural infrastructure is collapsing, services are disappearing, the local schools face closure. The reasons for giving up could fill another book, but we're not leaving.

As each year passes and the work on the farm grows along with my understanding of it, I seem to let go at the same time. Acknowledging that we don't really own it, that its life is so much greater and longer than ours, helps. It is a strange feeling of simultaneously being aware of the overwhelming scale of the job and finding the lightness in doing it. I meet the job face on and let go; as when two children play ring-a-ring-a-rosy, the grip must be tight but the movement swift and light.

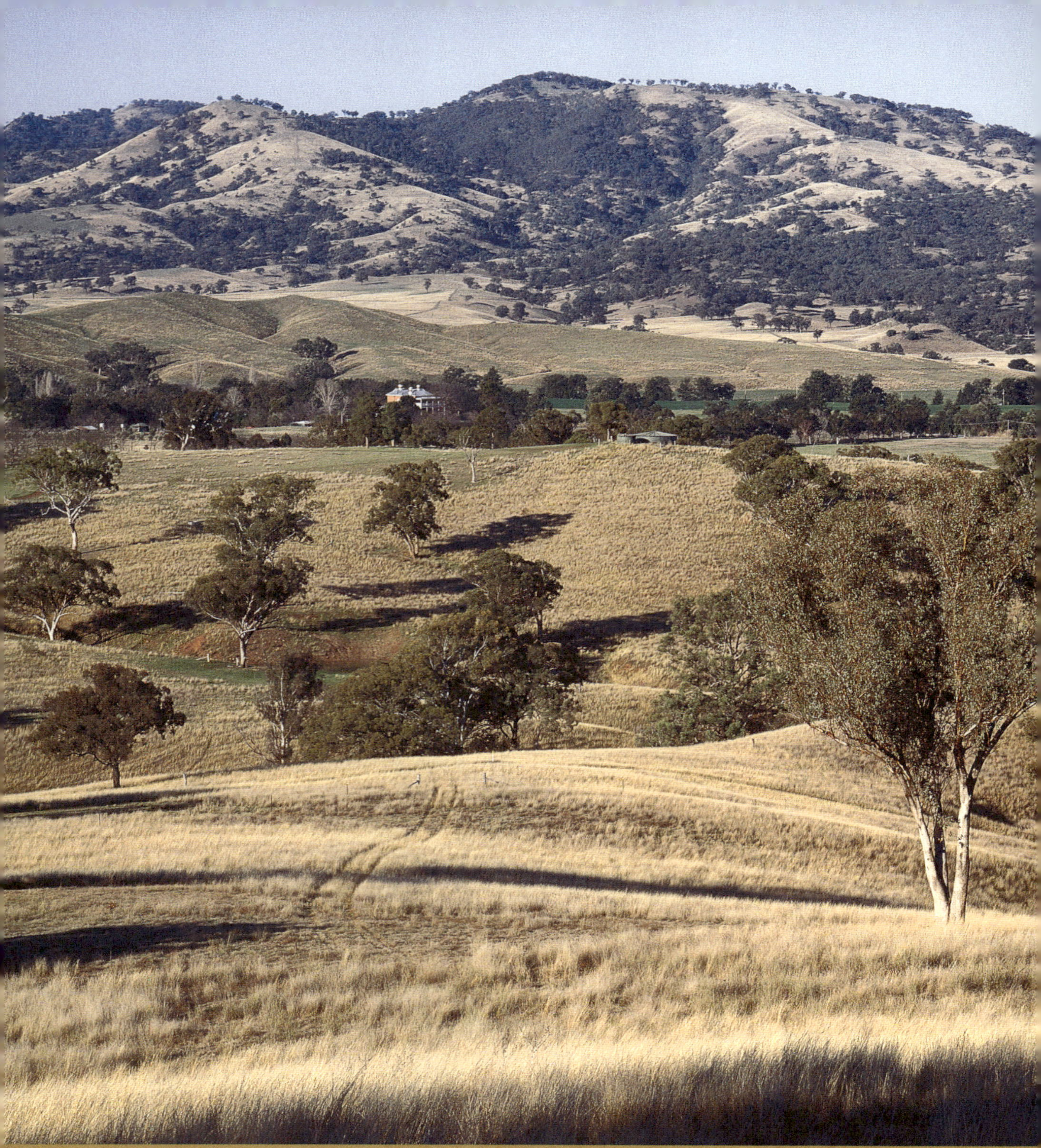

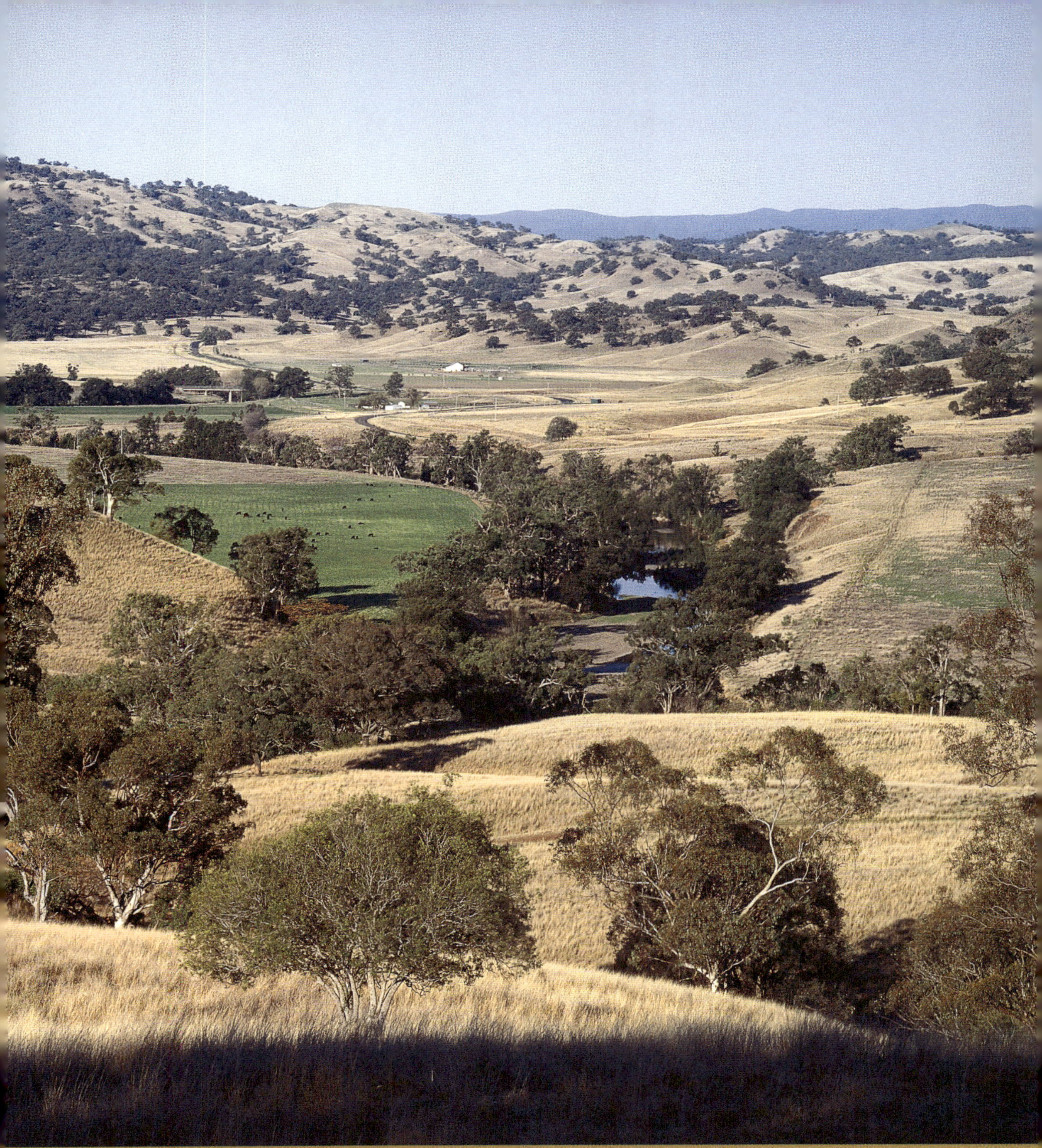

I frequently hear other farmers say they're developing their farms. It's becoming a liturgical chant. When I ask what they mean by development it usually involves poisoning the earth, eliminating native grasses, bulldozing and burning trees. It's exclusively about money. I too get excited by change, seek development and want our farm business to thrive. But in accordance with, not at the expense of, nature.

Development may be an ugly word but it's better than *growth*, which is now used incessantly by treasurers and bankers everywhere. We know that the economies of the world cannot grow unendingly without catastrophe. Can we defeat our faith in linear development? Deflect that arrow, remove that target? We're running out of earth, water, trees and time. Nor does using *sustainable* as an adjective for growth alleviate guilt. Sustainable growth is an illusion.

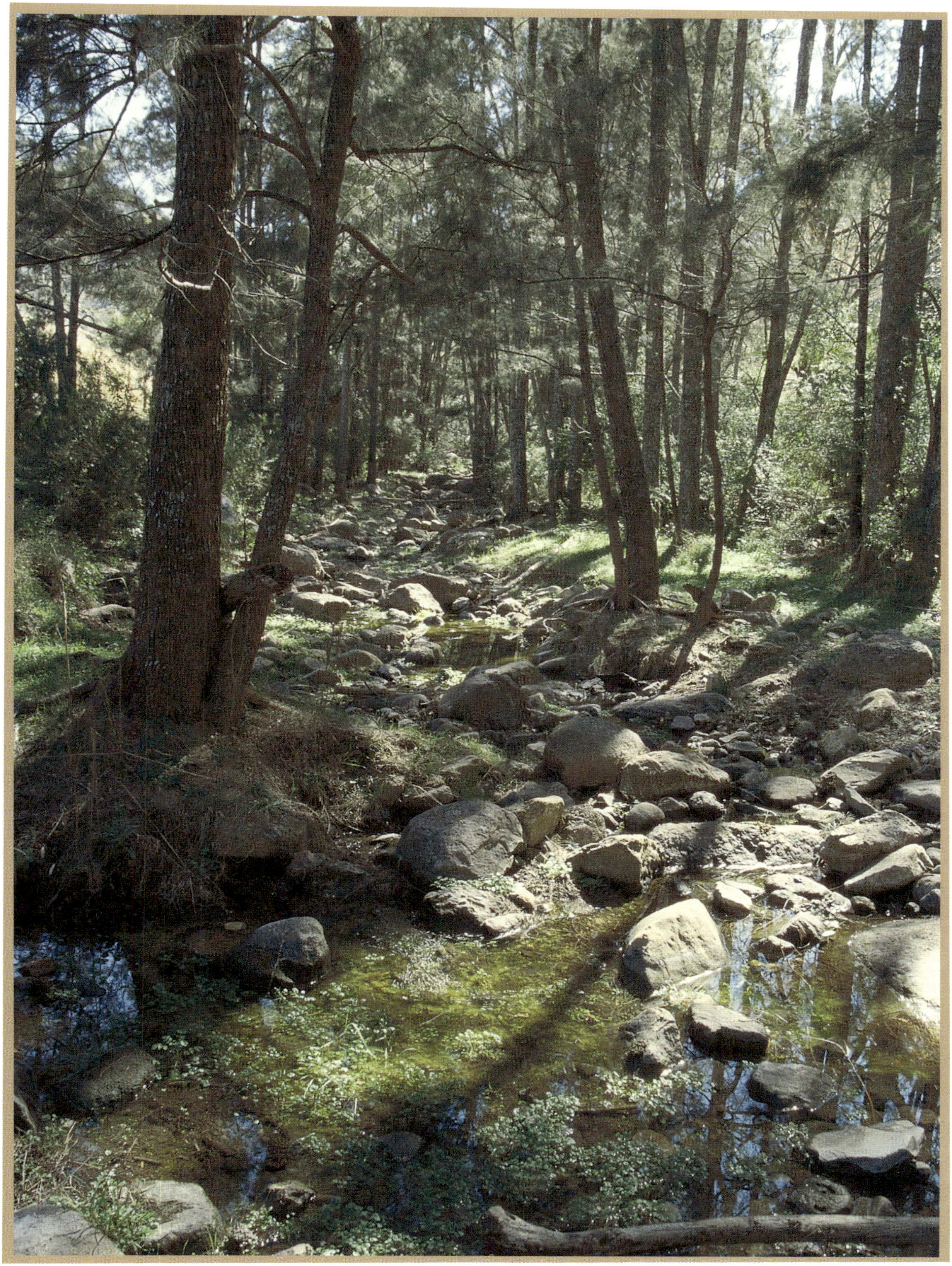

Autumn, and the olives begin to turn black. We test their ripeness by cutting open a fruit to see whether the flesh is hard or white-green and full of oil. Everywhere growers discuss their method of harvest and the appropriate timing. We are at once competitive and cooperative.

Getting the fruit from the trees has always been a major hurdle for the industry. Olives cling to their branches. Whereas a ripe raspberry seems to fall into your hand, olives are obstinate, and some varieties are notoriously difficult to remove. Though Koroneiki is recalcitrant, this didn't stop me planting a thousand of them when we extended the grove. While playing hard to get, the fruit has a distinct sweetness, in contrast to the bold bitterness of other varieties.

Contract harvesting is the answer for small producers in the valley. As the machines march through the rows, we walk behind them plucking the leftovers from the branches. Every olive must be picked lest fruit escape and the trees go feral, as they have in South Australia. A friend near Bordertown removes olive seedlings from her land every day, refugees from a neglected grove five kilometres away that have spread far and wide courtesy of crows and parrots. She's now organising pickers to harvest the fruit, for the sake of the environment and for fresh oil and pickling.

Trees have also escaped around here: Travis says that *Olea europaea* subsp. *cuspidata* (African olive) is the number-one environmental weed in the Hunter Valley. 'They totally modify the ecosystem,' he says emphatically, in case I don't quite get it. Their strong root system and bushy habit block out all competition and they take over. Apples and peaches have escaped too; we've found them in Elmswood's valleys and along the roadsides.

One day it will be up to the next owners — or Aurora — to assume responsibility for a vista of ageing trees. Otherwise they'll need to be grubbed out. Just as it's unacceptable to keep animals and neglect them, so is it to plant orchards and crops and forget about them. Native flora and fauna cannot cope with feral competitors.

The nest of a crested pigeon in the fork of an olive tree. The pigeon saw me coming, flicked up its tail and flew off. I hid, awaiting her return: she left the nest for nearly twenty minutes!

Another *Cymbidium canaliculatum*, tiger orchid – that's more than a hundred of them now. If I had the time for just walking I could find many others. I mark each one with a tag, hammering it into the trunk. The flowers, as with all orchids, are pretty, but it's the pods that impress the most. They're so much more flamboyant than the delicate yellow blossoms. That's the thing about Australian flora – the seeds often have greater appeal than the flowers, and they're around longer to admire

With Travis and two other enthusiastic botanists I'm bumping and bouncing along tracks near the Pages River in a four-wheel drive. They're speaking in a foreign language, and to make matters worse it's a dead one. They're talking in Latin, reminding me of Mass.

I feel like a tennis player who's been recruited for soccer. Though keen to play, I don't know the rules. When I ask for subtitles, my mild complaint is greeted with incredulity. To them, talking in the Latin of the botanist is entirely natural. They're not showing off. They're communicating, simply and directly.

I dig the notepad from my pocket and attempt to write down their foreign words. When phonetics fail I ask for correct spelling, all the time trying to connect the names with the plants I know. *Dodonaea viscosa*, hop bush. *Geijera parviflora*, wilga. *Bursaria spinosa*, blackthorn. *Hymenosporum flavum*, native frangipani. I'm colloquial, they're classical.

When we stop to collect seeds or stems for cuttings I look intently at the samples, determined to remember them. Not simply by the fine details – a rolled or folded leaf, a spike, a digitate seed head; not simply by the husk, the lemma, the palea – but by the totality of the plant. After all, you don't recognise a friend by a wrist, a toe or a fingernail.

This creates a personal, as opposed to a scholarly, order of beauty. So that weeks later I'm able to declare to one of my bouncing botanists that *Eucalyptus albens* is ugly. And he looks shocked that I should apply such a pejorative term to a tree. 'Let's be honest,' I demand. 'Compared to *noblis* it's, well, thwarted.' And yes, unpleasant. Next to *regnans* it's pitiful and pathetic. Although what other tree could possibly measure up to *regnans*?

I am condemned for being too discriminatory, a botanic bigot. 'There's beauty in everything, Patrice,' he tells me.

tank Pd
7/5/04   — open panicle

paddock love grass?
Eragrostis leptostachya

check ligule?
— should be a fringe of short hairs

high quality —
persists under heavy grazing.

Alisterus scapularis
Australian king-parrot

Fabled in legend, explorers' tales and literature as well as in astronomy, the Transit of Venus is under way. But few in our district seem to care.

Driving back from Aurora's singing lesson, we see no one outside staring at the heavens. At home I turn on the radio, but even the ABC seems unconcerned. We make the recommended viewing device from sheets of cardboard and use the pinhole to observe the sun. We're at one with Captain Cook in Tahiti but can't get the cardboard to work.

Leaping forward a few hundred years technologically, we try our computer, but it's too antique to connect to the live sites. (It's closer to Cook than to Gates.) Disappointed, Aurora retreats to do her homework while I go for a walk.

It is just on dusk and I can enjoy at least the idea of the Transit of Venus in a silence that the birds around me share. Everything is hushed. The experience is so eerie, so unprecedented, that I sense any Aboriginal astronomer millennia ago would have been alerted to the event by this stillness that verges on reverence.

The birds, like me, are witnesses to Venus.

They come back every July, just before Phillip's birthday. Two rufous whistlers. Are they the same two? While we'd only know for sure if we were to tag them, they look entirely at home as they flit from the Chinese elm outside the kitchen window to a banksia rose billowing over the garden shed. They sing and sing and sing, and after their performance move around to the north side of the house — to Phillip's library window, where the male taps its beak on its reflection in the glass.

At first it sounds as if it's applauding its own aria. Then the tapping becomes indignant and frustrated as the mirrored whistler pecks back, and the pecking escalates to the verge of madness. Should I intervene? Glue cloth over the glass? Prune back the branch? Consult an avian psychiatrist?

Why does the sun have to shine this way, shine exactly so? I tell myself I shouldn't worry about the bird, its beak, its brain. After all, my generation survived all that head-banging music at concerts.

A few days after the whistlers have left, peewees arrive and tap away at a different window, leaving the view of the river begrimed by their ceaseless attacks. Requiring dangerous manoeuvring with our longest ladder, cleaning makes the reflection even brighter and the peewees peck all the harder. Their mud nest, like a bowl thrown by a master potter, is perched in the *Eucalyptus noblis* twenty metres away. Every summer, dollar birds arrive and settle beside the empty nest. It is a very popular branch.

Come spring, usually after a generous rainfall, we'll be driving down the road towards the river and accidentally disturb fairy martins collecting mud. They explode indignantly into the air. Later, when they're calm again, we sneak beneath the bridge to see how their nests are going. Year by year the community has grown, reminding me of the mud towns in Morocco, or the Spanish pots Phillip has piled around the house.

No casual passer-by would guess that this ugly concrete bridge, built by the municipality in the 1970s, houses such a magnificent community. The birds are so busy building that if you take things quietly they'll ignore you. After a time, they seem to recognise our truck and don't even bother to fly away.

Nests of *Petrochelidon ariel* fairy martin

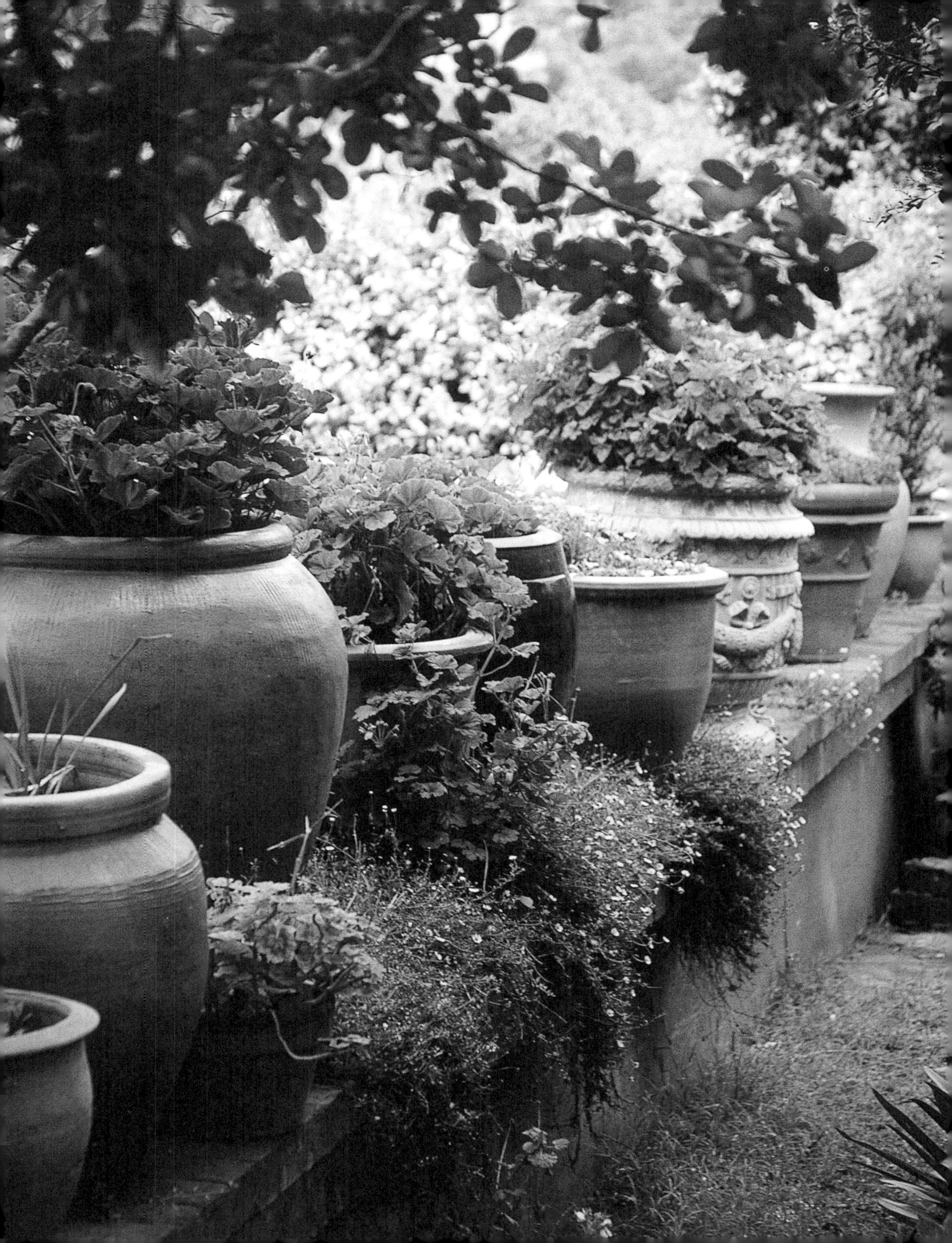

50
Ten Thousand Acres
A Love Story

Yet another drought. No rain drumming. The winds are less aggressive, more conversational. Algae gels the dwindling river. The grass looks as nibbled as fingernails. Dams have dried up and cattle trying desperately to get a final drink have died. Like troops at the Somme, they get bogged in the mud.

Our fences too, many of which have stood for eighty years, marching like soldiers up and down the hills, lie down and die. With their surreal collapse the farm becomes open, boundless.

Managing livestock is about containment and control, particularly during a drought. Now that becomes impossible.

When the drought is over I decide it's time to tackle an immense task: the removal of five kilometres of the most defeated fence and its replacement with a classic version made of fat strainers, stays, splits, and both barbed and smooth wire stretched as taut as strings on a violin.

Over the weeks we assemble hundreds of steel and wooden posts and choose a good-quality wire. I walk the hills with Glen, our fencer, discussing the best place to put gates. We find the half-buried remains of ancient post-and-rail fences, along with wreaths of rusty wire propped against tree stumps. And there's evidence of humpies and forgotten sheep yards in the groins of the hills.

Glen rests both big hands on a steel post, squints, and indicates a point on the next hill. His team has already placed the splits along the route but Glen's not happy with them. Too many are bowed. I quite like their roughness, the way they evoke the trees they were just a few weeks ago. But Glen protests. 'You shouldn't see anything bulging when you look from one strainer to the next.' We compromise, with the least straight splits relegated to the fence around my vegie garden.

As we pace out the fenceline, Glen proudly tells me that his dog has tracked down some of our cows that had gone missing. Then, almost shyly, he asks if he can go pigging on the farm. This is not a job for his cattle dog but for Oakey, a pig dog, and the pigs they hunt down will end up in a cool room in Murrurundi, prior to being exported to Germany.

Whereas Travis and I walk the hills looking at leaves and grasses, Glen watches for the prints of trotters and the digging of snouts. As pig numbers build, they move closer and closer to the homestead, excavating in the olive grove and threatening my beehives. I'm more than happy for Glen to go pigging.

'I use glow bars as collars,' he explains, producing a luminous strap from his pocket. 'So if you see lights in the night, don't freak, okay?'

Pig dogs wearing jewellery! I laugh at the thought of Glen and the bedecked Oakey tiptoeing through the scrub.

After dinner I see the lights of the truck and the glow of Oakey's collar. When Glen arrives at the house he's lugging a dead pig over his shoulder, and we hang it in the meat shed. In the morning we skin it — we haven't the gear to remove the bristles, so there'll be no crackling. We drag out the band saw, cut the boar into portions and rest it in the fridge. This will set the meat and make the detailed butchering easier. We'll make sausages out of the trimmings.

Later, I'll make a pork marinade. Onion, honey, thyme, white vinegar, tomatoes, Worcestershire sauce, stock. Cook it up, cool it down, pour it over pieces of spare rib in a baking dish and leave it in the fridge. After four hours it'll be ready to cook — for one hour in a moderate oven. The marinade will caramelise and the pork will have a soft, sweet, sticky coating.

Ignore allegations about wild pigs being riddled with worms; we see no evidence of this. These free-range animals are strong and healthy. It's a shame to see them shipped off to Europe, especially when Australians are forced to eat factory-farmed pork that's been fed muck and injected with drugs.

Feral pigs have a culinary reputation not unlike that of carp. Both get bad press. At best, carp is minced up for fertiliser, yet here's a large and plentiful fish that can be made into delicious meals. The problem isn't cooking them, it's catching them. We're told to put corn on the hook, but that works no better than worms or yabbies. While they eat everything else, they never take the bait.

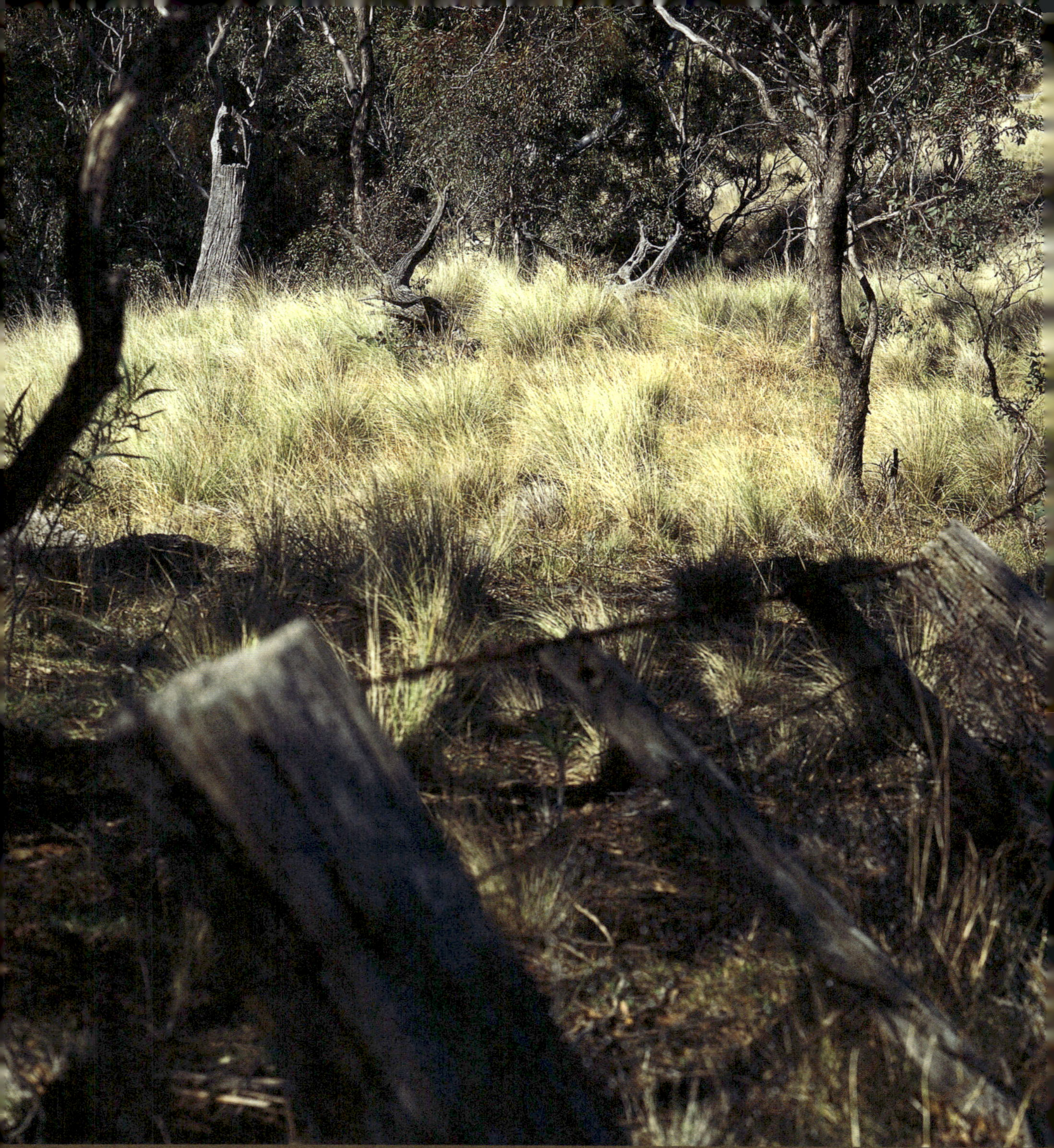

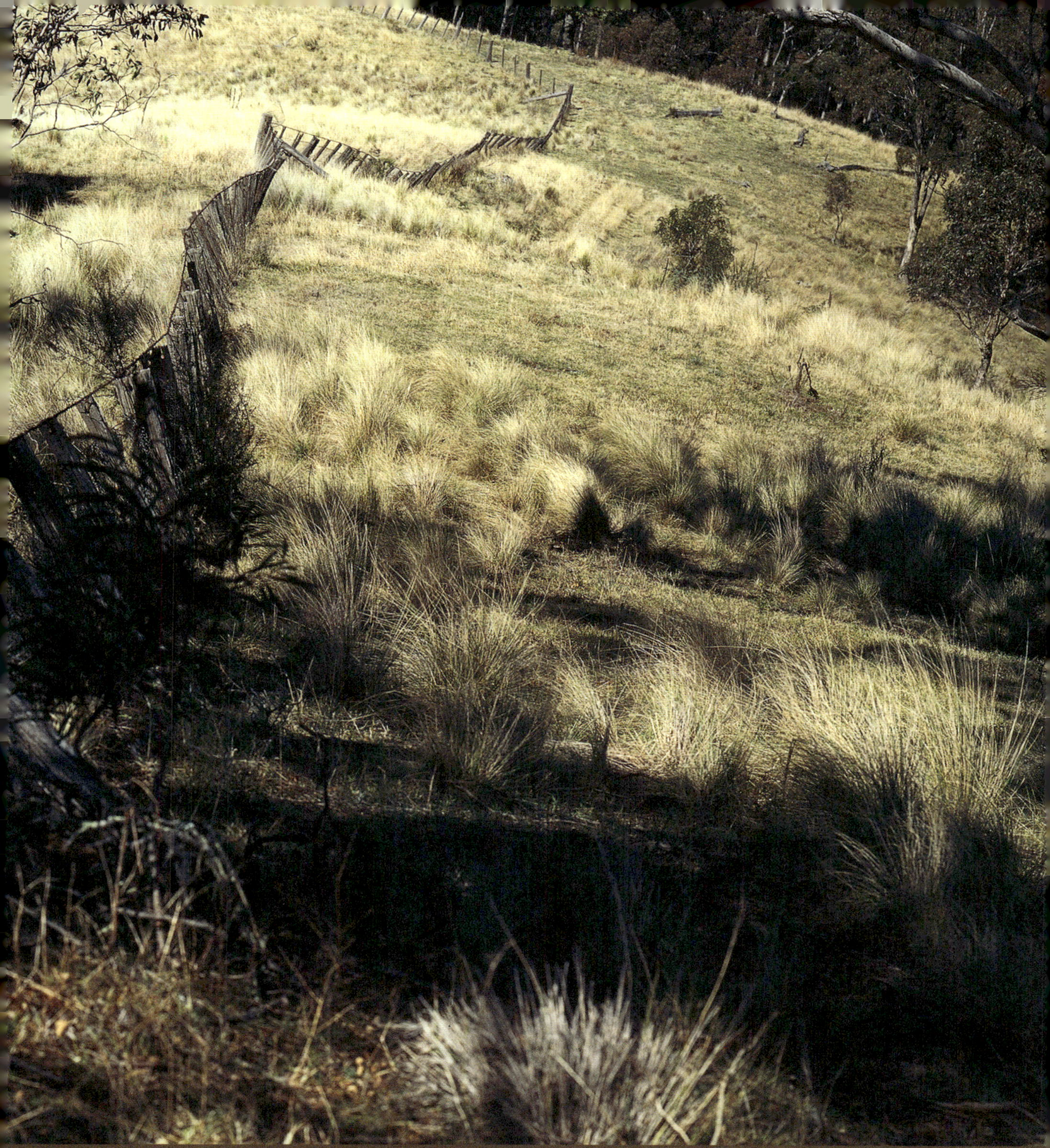

Come September, noisy miners return to squabble in the shrubbery. They're fighting over real estate and development opportunities. No one likes these birds, but I'm impressed by their finesse in nest-building — the way they pick up old she-oak needles and flick them around, judging each follicle's strength and length before adding it to the structure. Not to be confused with the even noisier and more disruptive common myna, at least miners have the virtue of being native.

Casuarina cunninghamiana subsp. *cunninghamiana* river she-oak

There's a lot of death on farms. It's inescapable. Each year we send animals off to the abattoir and their flesh finishes up in people's ovens, on their plates. Or the bones are ground into fertiliser to keep the cycles of growth going in our gardens and farms.

I watch the swoop of a hawk as it scoops up a lizard, hear the death squeals of rabbits in the jaws of scuttling foxes. We used to lose many chooks to foxes, but not since the arrival of our black-dappled Muscovy drake, a breed that doesn't quack so much as rasp.

Although Muscovies can fly, ours prefers a sedentary existence. He lives in the chicken coop with the rooster, watching over the Ancona, Langshan, Leghorn, Isa Brown and Sussex hens, bobbing his neck like one of those plastic dogs on a car dashboard. If I don't put enough feed out he whispers angrily and follows me around until I do. I'd pick him up for a cuddle if it weren't for his huge toenails, forever in need of a pedicure. They'd curve right around my fingers. The foxes must be intimidated too, because they keep their distance.

Aurora began collecting cattle skulls when she was two. She and her dad would go out on hunts. 'There's another one!' And she'd climb down from the four-wheel bike to retrieve it, carrying it back in triumph. Some were as big as she was. Unable to lift them, she'd drag them along using a horn as a handle.

Here the skull of an expensive bull that died mysteriously, there a cow that fell off a cliff. Another that had been dead perhaps eighty years. In their own strange way they were compelling, beautiful, striking sculptures.

Upon their return, father and daughter would examine each item with solemn fascination, as if preparing for a funeral ceremony. Every skull had some distinctive feature: a wider forehead, a brighter whiteness, a more spectacular pair of horns.

We're always coming across the bleached bones of the long dead in the long grass. Not only cattle and sheep but native animals, including one inexplicable skeleton that turned out to be a platypus. Once, we found something that had been half eaten by — what? A goanna, a wild dog, a pig, and then maggots? We thought it might have been the corpse of a rare quoll but finally identified it as a koala from its massive, tree-climbing claws.

From time to time Aurora and Phillip would augment their cattle skulls with the frail relics of a kangaroo, rabbit or bird. The front verandah came increasingly under siege. It already accommodated some of Phillip's stranger collectibles — a vividly painted Sicilian wedding cart; a more spectacular four-wheeler with a lattice dome, once drawn by camels in Rajasthan; and a bark canoe from a trip to the Maningrida Aboriginal community on the Arafura Sea. Now all of them were loaded with a cargo of skulls.

One weekend while I was away, Phillip and Aurora heaped all the skulls onto the back of the truck and drove to Dam Paddock for a ceremony, to

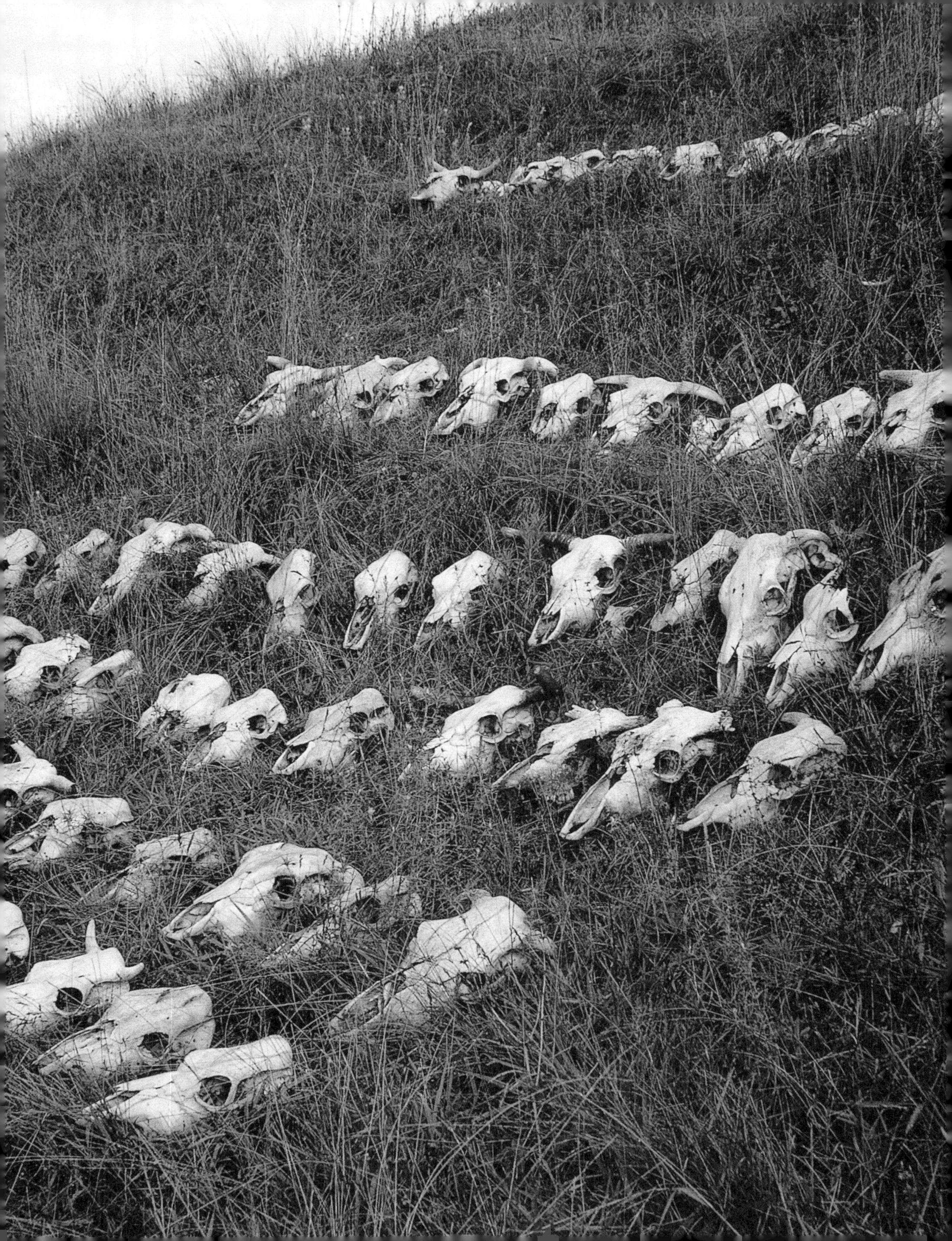

which I was not invited. A pity; I'd like to have attended. When I got home I saw something strange on the hillside. It looked like row upon row of cockatoos, taking a break from destroying the homestead's *Celtis* trees. I reached for my camera to record this extraordinary moment — scores of the big white birds in a unique configuration — and when I focused the lens I saw the most macabre arrangement: an open-air graveyard bearing witness to the hardships of a century of cattle breeding.

The hillside had an amphitheatre, made not in the style of the ancient Greeks but by previous owners' sheep. Sheep move across hills in such a way that generations later they've created horizontal paths, like the tiers you see in Bali. Phillip and Aurora had arranged their skulls, at least a hundred in all, to sit gazing with empty sockets at an imaginary theatrical event. There were skulls in the stalls, in the dress circle, in the boxes.

Phillip had a range of names for the amphitheatre. The Bullshoi. Cowvent Garden. The Bullvoir Street Theatre. Cownegie Hall. Her Moojersties. Aurora announced there'd be performances of Moozart, Beethoofen, Gustav Moolah's second symphony, and a Tchaicowsky bullet.

Months later, when live cattle returned to Dam Paddock, they showed complete disrespect for the resting place of the deceased and walked right through the skulls, sending them tumbling into a gully, the bones falling like skittles in a circus sideshow.

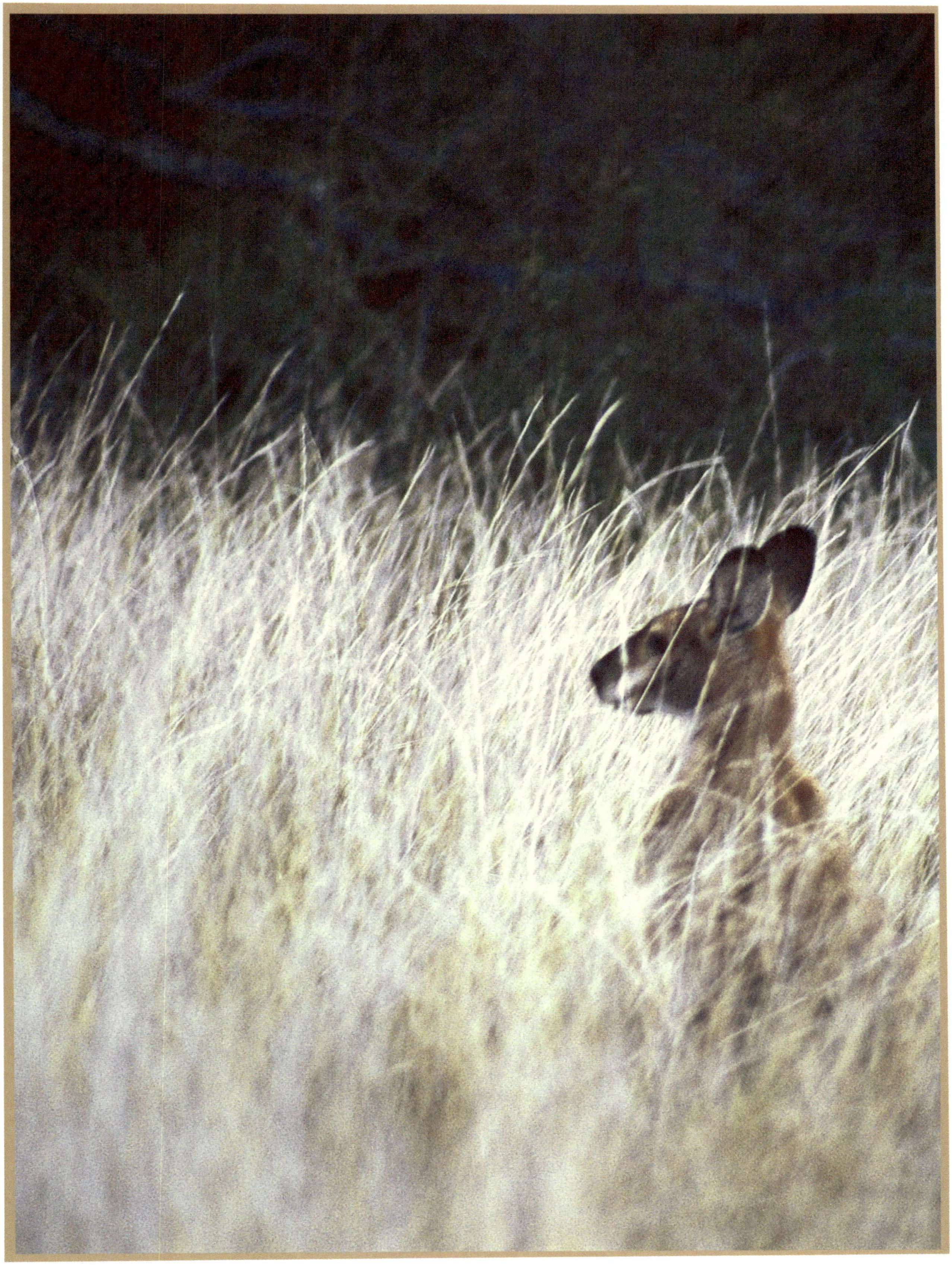

Our rosewood tree, *Alectryon oleifolius* subsp. *elongatus*, hangs onto life with such determination. Every time it defoliates I'm about to post a death notice. Ants climb all over it, seeking scale. Up close it's like a busy train station, with ant tracks going every which way, back and forth to their huge mound by the trunk. On mornings when it's clouded in a hangover of fog, the tree makes me smile. It's so brave the way it struggles for life

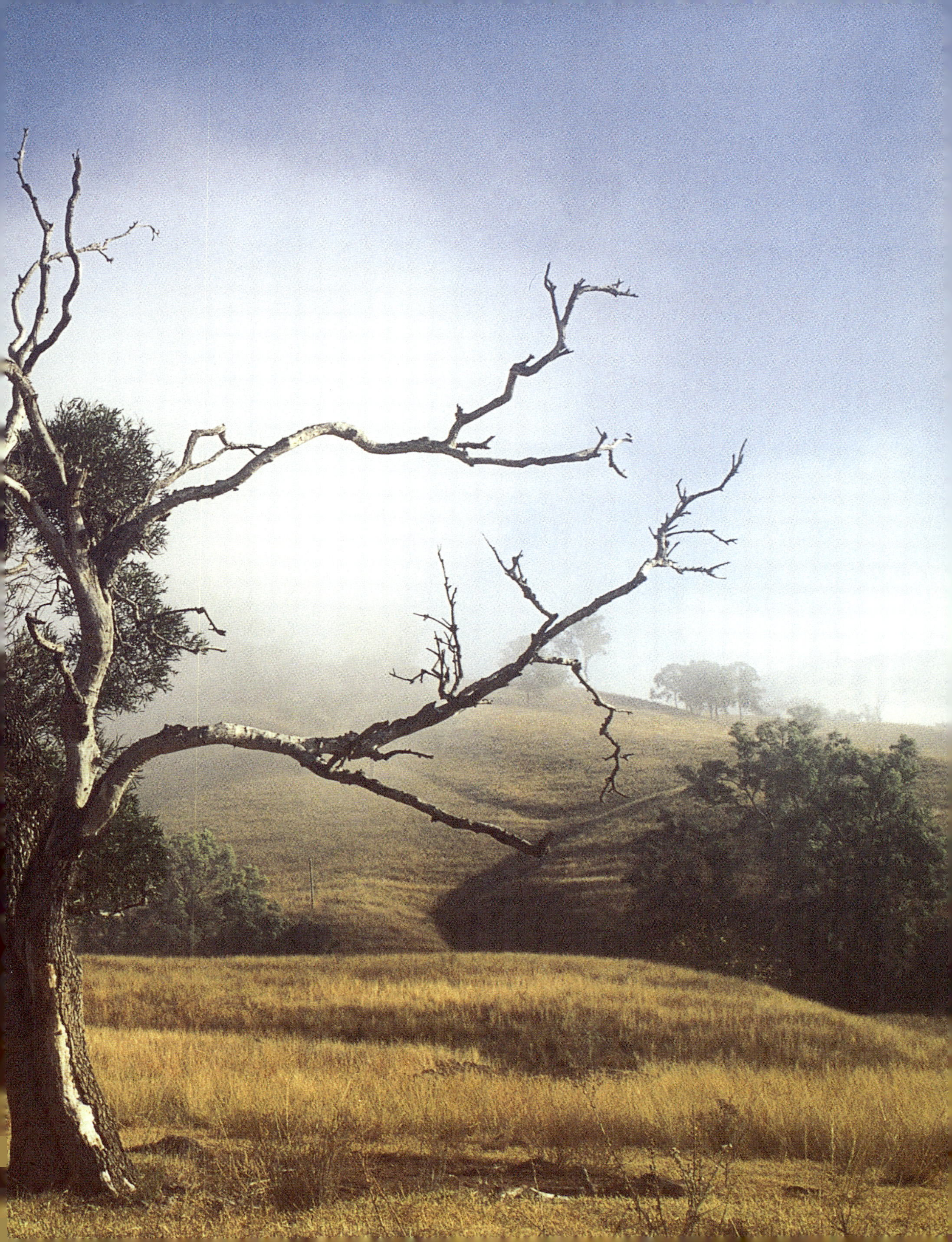

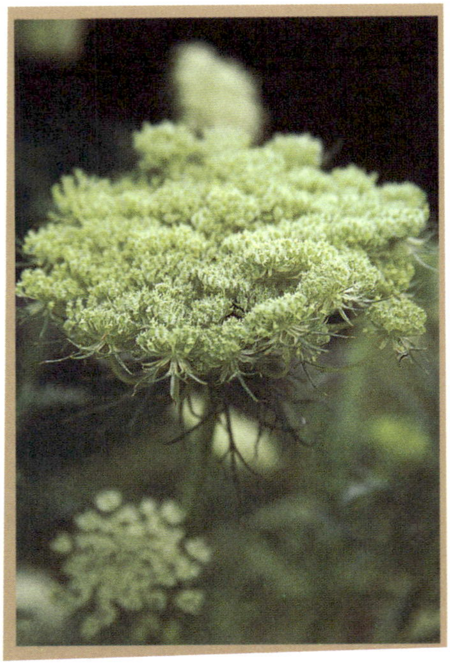
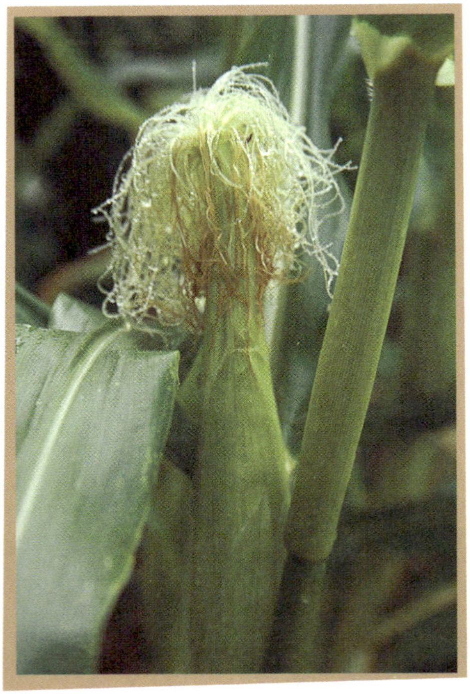
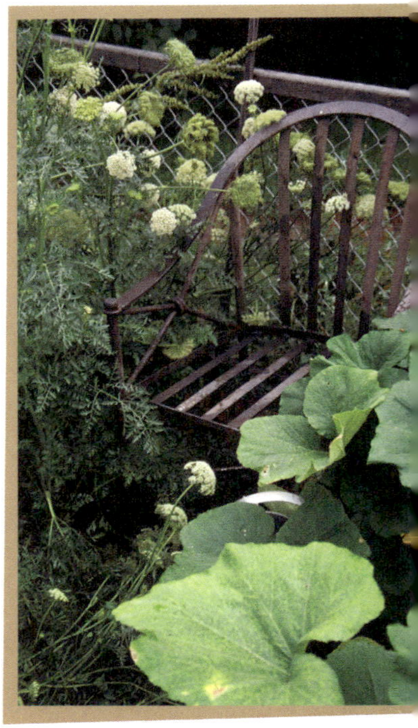
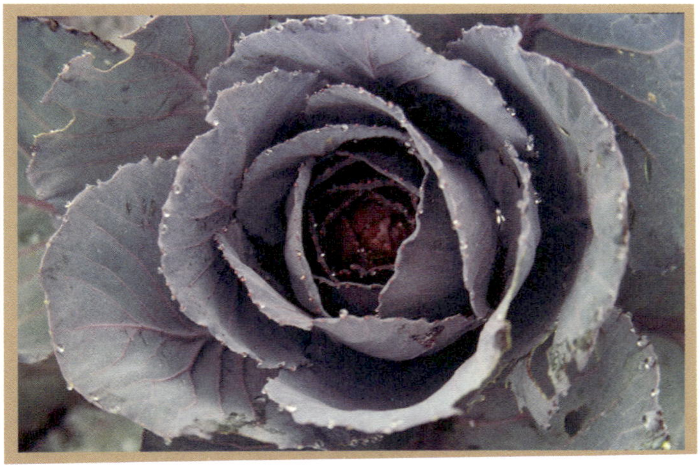
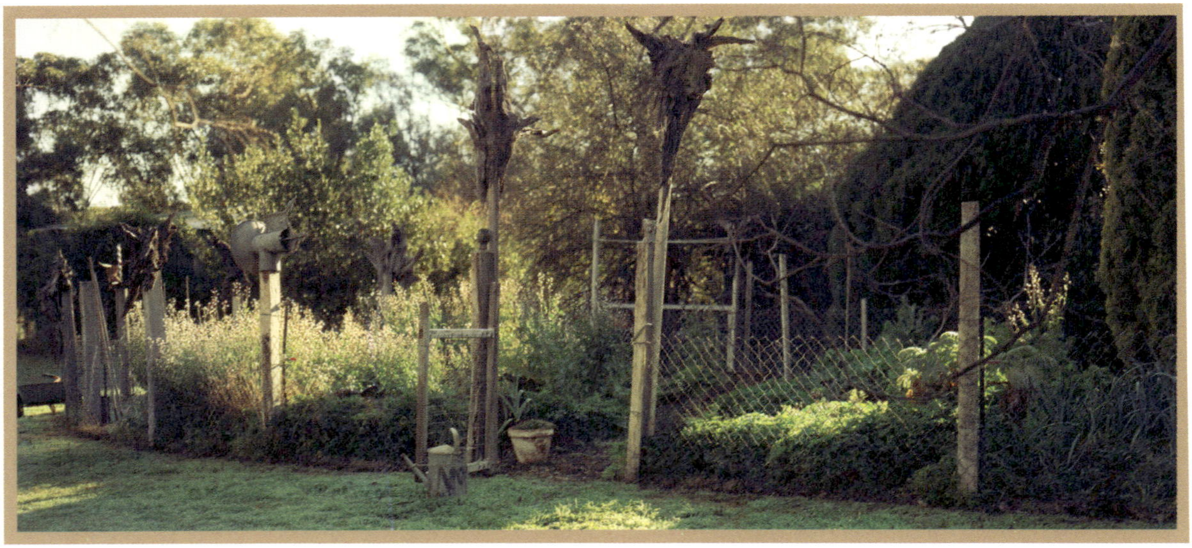

On Saturday mornings I get up at least two hours before Phillip, make a mug of green tea and take it to my vegetable garden. When he finally stirs, Phillip collects the freshest eggs and then drives to Gundy for the newspapers while I prepare an omelette. Aurora sets the table on the verandah, I pour boiling water into a Wedgwood teapot, and we divide the papers between us. Phillip says Saturday morning's omelette is his favourite meal of the week.

I sweeten the omelette with leeks in spring and onions in winter, and garnish it with parsley directly descended from my childhood garden. We have quite a few dynastic plants around us. As well as Mum's herbs, there are violets from Violet Grove, East Kew, in Melbourne, where Phillip's grandfather had his little flower farm. There are pots of the drought-hardy *Billbergia nutans* (vase plant), its pink flowers outlined in blue, from Phillip's English grandmother's garden in Warrnambool. I've explained the history of the plants to Aurora so that she can tell her kids about them.

After breakfast it's time to check the cattle, our other Saturday-morning ritual. I climb onto the four-wheel bike behind Phillip and we head off, dogs bounding beside us. I want to see how the grass is faring, where the herds are camping, how the water's lasting.

Many of our paddocks are huge and thickly timbered and I ask Phillip to drop me off so I can walk. We arrange to meet on a distant hill, after he's detoured to chainsaw a tree that's fallen in the latest storm. I hear the scream of the saw as my binoculars reveal fifteen, sixteen, seventeen head of cattle on a ridge. They shouldn't be there. Rummaging through my knapsack, I find the electric-fence tester, hook it onto the wire and watch the glowing readout. It confirms that seven thousand kilowatts are pulsing strongly from the shed by the river three kilometres away.

There are nine cows and eight big calves on the ridge. They're too far away to identify, so I head off to investigate — and here's the answer. A gate lies on the ground like a welcome mat and the whole herd is about to accept the invitation. A bull can lift a heavy gate from the hinge with its nose, yet it's far too heavy for me. The best I can manage is to prop it up. That will have to do for the time being. The old saying 'a bull at a gate' rings true.

Gavin Prescott arrived at Elmswood four years ago. We only expected him to stay a few months but he's still with us. As he often reminds me, he's a 'Suffuckian' — from Suffolk, England — where his father taught him, among many skills, to fix farm machinery. He's never happier than when he has his head under a tractor's bonnet, or his arms deep in the guts of a baling machine. If there's anything he can't fix we've yet to find it.

Where others who've worked on the farm got vertigo on the first rung, Gavin climbs ladders like a circus performer, and tightropes across the homestead's loftiest peaks, entirely certain of his balance. When he makes a new gate for the vegie garden he'll shape the latch to fit the curvature of the hand. If it's a wooden box he's making the nails are placed with perfect symmetry, and when he's painting a window frame there won't be a single drop of wasted paint.

This neatness, learnt in Suffolk, was honed in a subsequent life working and living on boats, where life depends on attention to detail, and on ritual. Now Gavin is forever threatening to sail away on his 63-foot catamaran. We've been a part of his adventure as he's built it and we all want to go along for the ride. We'd trust him completely, even if caught in a perfect storm. A man of the land, a man of the sea. Ploughing through waves or through the earth in the front paddock.

Today I've scheduled another argument with him, this time about the new olive planting. Picual or Coratina? There are always arguments about ploughing. Should we be deep-ripping between the olive rows? Should we plant annual crops or allow the native grasses to take hold? Can we do both? Either way, Gavin will want to sow neat, perfect rows. But a lot of our gear is old and inappropriate, so he'll improvise, sowing buckwheat — as an experiment in biological control — with a spreader, for example. Despite his best efforts, it lands willy-nilly, and when clouds smudge the hillside and the air moistens just short of rain, ants become enthusiastic and spread out, eating the seed.

Nonetheless weeks later, after a duststorm and a rainstorm, the buckwheat jumps out of the ground. Butterflies, spiders and sundry bugs rejoice in it.

If you pick buckwheat it wilts before you have time to admire it. The big heart-shaped leaves, as beautiful as a shamrock's, make me want to press them as keepsakes. But the flowers, despite their delicacy, smell like old shoes.

The strengthening sun haloes the mountain. A breeze stirs the windmills into complaining rotation. They're not working very well. Gavin will later tether them with wire until we can tackle the repairs. In any case, the water table has dropped so far that the windmills are sucking air.

Beside a dry creek, the rusting blades are sheltered by two soft, bosom-like hills. Windmills, as do the sails of a boat, need to be out in the wind. In good times the blades blur like the propeller of a plane and water gushes into the dams. So why was the windmill put by the creek in the first place? I wonder. There's little in the way of wind, and these days not much in the way of water.

The ever-increasing dryness is a part of disturbing, untrustworthy new weather patterns. We no longer know the weather. Once we had fogs during the short days of winter, and again at nightfall. Now they're a fading memory, an infrequent pleasure. Clouds rarely loiter; they promise rain but hurry by. We've become used to disappointment.

Because it's so dry we wean calves early, giving their mothers an extended rest before the next calf is born. The separation of cows from their progeny takes place in the cattle yards and is a time of chaos and emotion, our sadness appeased by the excitement of the animals as they rush to take hay from our hands.

No need to dig the vegies out, because summer is baking them while they're still in the earth. Potatoes are on the way to being crisps, and tomatoes are sun-dried on their stems. January hammers my vegetable garden with ten consecutive days over forty degrees. And where are we at the time? In the world's most watery city, Venice. We return somewhat guiltily to Gavin's tales of woe and wilting. The last of the corn, beans, potatoes and zucchini are dehydrated or dead. The surviving tomatoes are in intensive care, but within days I switch off the life support and pull out armloads for the chooks. No green tomatoes for pasta this autumn. Only rainwater, enriched with life from the air, can create the best quality vegetables. We plant new seeds.

One February after a dry time, native daisies and bluebells burst forth like never before and whole hillsides become wildflower meadows.

I meet Aurora at the school bus stop and take her to see them. It could be a once-in-a-lifetime opportunity. I dig in the glovebox for the camera.

As the sun lowers and the light begins to fade, the mauves and yellows intensify. *Wahlenbergia gracilis*, sprawling bluebell; and *Calotis lappulacea*, yellow burr-daisy. We kneel among them, picking some to take home. For all their visual delicacy, they're tough and hard to pick. We finish up pulling them out by the roots.

Calotis lappulacea
yellow burr-daisy

1/03

Wahlenbergia

Blue Bell

There's a stench in the hallway. Bats in the attic. It's so hot outside that we shut and shutter the house. This intensifies the smell, and by bedtime we fear asphyxiation.

Before dawn I lie in bed and watch scores of bats returning from their nocturnal adventures. A flock of Draculas seeking the safety of the coffin. Their silhouettes funnel into single file as they disappear into a fissure between the roof and the eaves. Once inside, they'll organise themselves into dangles, like washing on a line. Fine, except for the droppings on the verandah and the stink in the ceiling.

We scrape up a bucket of poo each week, and fear for the weight of it building above the bedroom ceiling. I take a torch and aim the beam into the aperture that grants them entry. In confusion and panic, they collide with each other and hit the roof. Their aerobatics are so eccentric I feel sorry for them and switch off the torch. By the time I've had a shower, the bats are asleep and silent. But they've left a baby behind, dead on the verandah. I pick it up and gently open a tiny wing. It's as ancient as a pterodactyl's. Aurora, holding its outstretched wings, says, 'It looks like Jesus on the cross.'

Later, I have it identified: a southern freetail bat, a species of *Mormopterus*. Most of the tiny creatures we find in the bathroom, that swoop in front of the TV during *Kath and Kim*, that drown in the lily pond, that get into our clothing, and even into Aurora's bed, are *Nyctophilus geoffroyi*. Usually roosting in tree hollows, they're commonly known as the lesser long-eared bat. The two species keep their distance in different parts of the roof.

When we try to rescue them, to pick them up and throw them into the night, there's a danger of being nipped on a finger, and they carry a disease distantly related to rabies. Phillip and Gavin climb over the roof, trying to block their entrance, but in no time at all they find another one.

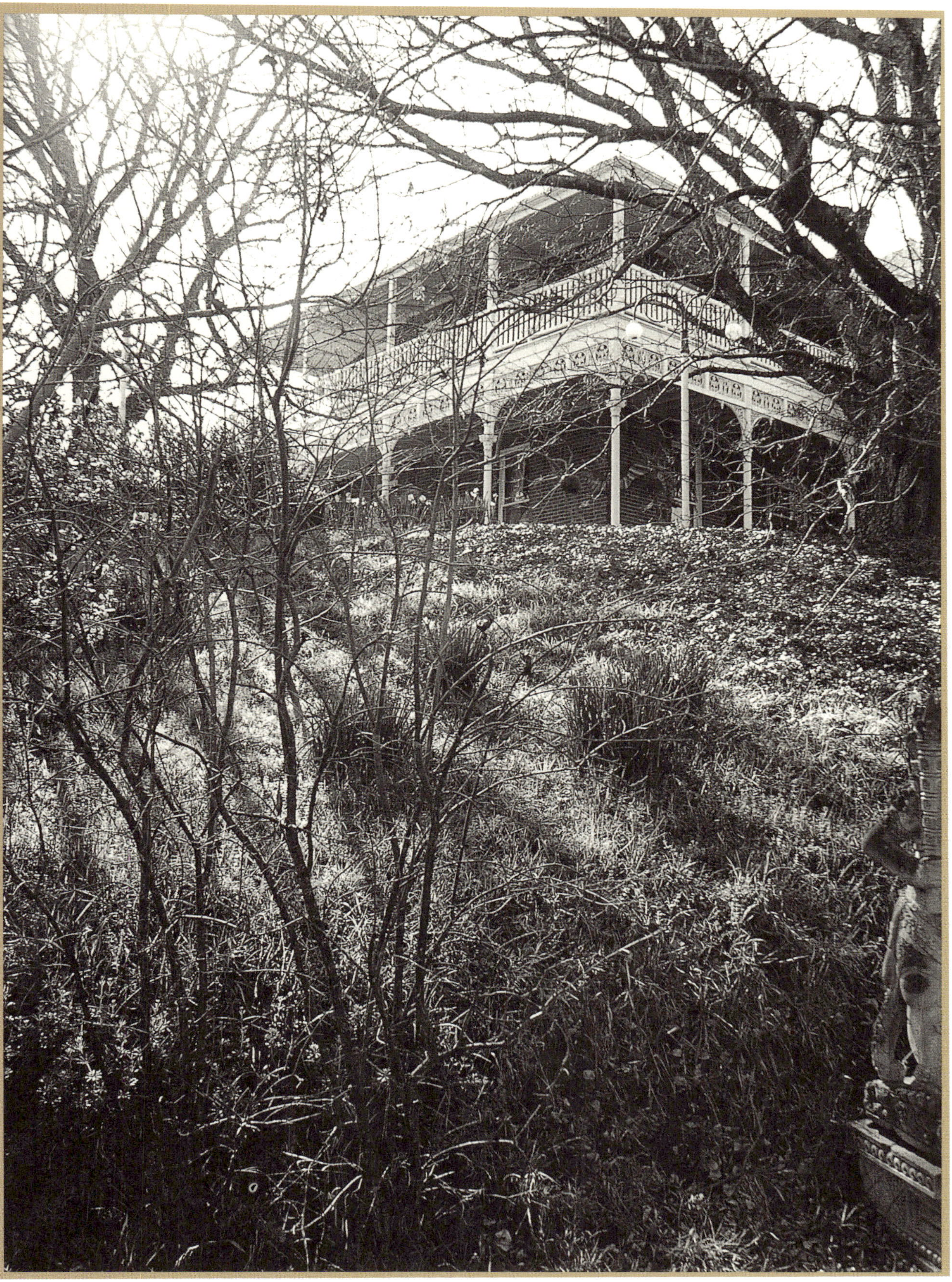

The white cedars, *Melia azedarach*, along Kewell Creek are in full mauve flower. These natives are our own European-style tree, deliciously deciduous and decorated with golden berries that linger through winter. When I was growing up in Adelaide's Kurralta Park they were the street trees, until one hot afternoon when they fell victim to a municipal death sentence. Along came a truck with burly blokes who cut them down one by one.

My mother, unpersuaded by the council's argument that they were dangerous, hardly spoke for days. 'Why?' she kept asking an empty room. Because apparently people were slipping on the berries. Defiantly she retained her own personal white cedar in our front yard, its branches dangling over the footpath. Having once had its limbs entwined with those of the council trees, it would forever after be lonely.

Melia azedarach
white cedar

Travis calls in to return some books on his way to work. He's wearing a new white shirt and its creases disappoint. 'I don't iron and this is meant to be non-iron, and when I put my jumper on all this fluff came off.' He picks at it despairingly. Poor Travis has to look his best these days as he's been evicted from the public service in cutbacks and endless staff restructures. Now he has to fight for survival in the private sector.

I show him a print of a 1792 botanical illustration mysteriously entitled 'The Cumberland Tree', a broadleaved specimen that looks like *Clerodendrum tomentosum* with its clusters of star-shaped flowers that open to red calyces. *C. tomentosum* grows along Kewell Creek. Travis agrees with my identification but has never heard of a Cumberland tree.

The artist, John Doody, was a convict servant to Colonel William Paterson, and his painting is a beautiful rendition of a spectacular tree. You can sense how impressed Doody was when he saw it more than two hundred years ago.

Given the deepening water crisis, *Clerodendrum* deserves greater respect and a stronger marketing effort. Gardeners are quite rightly admonished for growing thirsty broadleaved plants, but *C. tomentosum* is a native broadleaf, like sweet pittosporum (*Pittosporum undulatum*). They grow among our she-oaks, enjoying the dappled light, but their roots also dig deep into dry outcrops to become lonely specimens on windswept hillsides. They're the most southern *Clerodendrum* in Australia and are not big drinkers.

Only the flower of *Hymenosporum flavum*, the native frangipani, is as beautiful, but frangipanis need a lot of water. I find them growing in Elmswood's gullies, protected in summer by deep shadows and drawing moisture that leaches from the hillsides.

Clerodendrum tomentosum
hairy clerodendrum

There are many licks around the property and all the animals, wild and domestic, use them. Many tongues can gradually turn a cliff face into a cavern.

The interior of the cave — honey-hued, with a suggestion of mauve glimmering at the edge — turns grey as it recedes, one mineral giving way to another. If you approach slowly and quietly you'll find wallabies lounging on their elbows, ears up, noses coloured as if by cosmetics. They look like rajahs luxuriating in an Indian miniature.

Aurora scrapes some of the rock away, puts the powder into tiny jam jars retrieved from motel breakfast trays, and uses it as body paint.

Macropus rufogriseus
red-necked wallaby

The ceiling of the Sistine Chapel is a masterpiece, as are Picasso's *Guernica* and Goya's erotic *Naked Maya*, but for me a more significant artwork is a simple painting of a tuft of grass. Painted five hundred years ago by Albrecht Dürer, it deserves its grandiloquent title of *The Great Piece of Turf* because history records no earlier example of such a picture — of an artist focusing on something as inconsequential as a tuft growing in a meadow.

Dürer insisted that only by knowing nature could one know truth. So he made that piece of turf his turf, and half a millennium later the purity of his intentions seems all the clearer.

Dürer's times were marked by a stark division between rich and poor (echoed in our own, wherein the dieting and the starving share the world). Peasants were driven from their farms to become refugees on the road, searching for food. Northern Europe was plagued by Black Death, syphilis and famine. A few could afford to pursue beauty, most had to contend with a dangerous, desperate world. While Michelangelo and da Vinci were lifting their eyes to the heavens, their dreams and ideals focusing on transcendent beauty, Dürer saw loveliness in the earth.

We need our Dürers more than ever today — those who find truth in nature. A good farm is about truth. A good farmer can be an artist too.

Phillip — and no one knows me better — says, 'The farm's been good for you.' Certainly it has kept other worlds at a distance; it can obliterate their sound and fury. And it makes me conscious of my limitations. Every day, when I get up at dawn, I'm determined to try.

Dürer's *The Great Piece of Turf*

'It came from outer space,' says Phillip, beaming with excitement as he holds a grey rock in his open hands, as if it's John the Baptist's head on a plate. 'I think it's a meteorite.'

It certainly looks extraterrestrial, its structure aerated with holes. If it hasn't fallen from the heavens, it's pumice from Vesuvius, or something hacked from a furnace at BHP.

Phillip found the rock lying on the grass in Windmill South Paddock. 'I've never seen anything like it,' he says. 'It wasn't buried. It was just sitting there.'

The size of a turkey, it looks massively heavy but is as light as fairy floss. We agree that one of us will take it to Sydney for geological examination. The diagnosis, when it comes, is that it's a lump of iron oxide 'revealed by erosion'. Phillip pooh-poohs this, telling Aurora it's from the planet Krypton.

I once longed for a sweet, sentimental, romantic, John Glover-type garden, full of seasonal diversity and abundance, but I have to admit I've lost the plot in my garden at Elmswood. I still weed it occasionally, but have allowed it to take on a life of its own.

In the beginning I felt like a conductor with a floral orchestra, and there was much music to be had from plants given by friends: viburnum, Jerusalem sage, obedient plants, berberis, lemon balm, mints, crab apple, flax, hellebores, evening primrose, wormwood. For a time they played in tune, reminding me of the friendships that gave rise to them, but I need to rescore the garden if it's to become a fully-fledged symphony again. I must consider changing weather patterns, maintenance and, most important of all, how the garden will look in ten or twenty years. Can I possibly find the time?

Garden plants can be prima donnas; they thrive on flattery as much as cow manure. Pentstemons were to be the dominant perennial in my garden, on the basis that they're as hardy as they are pretty. But they flower, fall, demand watering, and, as soon as I turn my back, wither. Needing an accompaniment for the roses, which I still manage to prune, and with less and less time for floral tantrums, I yank the pentstemons out and change my allegiance to salvias and a great jumble of pots, from the miniature to the massive, brimming with succulents and geraniums. Now my failures are forgotten and the successes don't seem to need me any more.

The olive grove from our
bedroom window, with gaps
in the rows where we lost
trees to frost

One of the reasons for planting the olive grove was a dawning realisation that its chorus lines of trees would be more manageable for an ageing woman. Certainly it will be less taxing than sorting hundreds of head of cattle in the yards. Olive trees don't butt, kick or bellow.

The grove is more robust now than in its early days. Seedlings too frail or susceptible to frost have been replaced with sturdier varieties. Their roots go deeper in the heavy soil and take the battering of stronger winds. With my trees more at home, I'm no longer fearful of bitter chills or week after week of scorching heat.

But I remain fond of our cattle. After years of us culling, and buying better bulls, our herd has grown into a fraternity that knows Elmswood. Except for a few that have the Naughty Gene, they've become ever easier to muster, looking forward to the next paddock and knowing where to find the dams and troughs.

The ancient-Greek physician Hippocrates is credited with a profound piece of advice to all who would practise medicine: *Primum non nocere* – First, do no harm. Good farmers practise medicine with the land entrusted to them, and it was this approach that drew me to biodynamic agriculture. Of all the methods of farming, biodynamics does the least harm, and it can do a great deal of good.

Biodynamic agriculture is not as popular as it should be. While its benefits are widely acknowledged, many new farmers are oncerned that it replaces a reliance on chemicals with a dependence on machinery.

When we arrived at Elmswood in 1987 many farmers in the district had their own equipment for making hay or silage; they had lumbering machines to sow new pastures and crops. But increasingly, in an attempt to cut costs, such equipment was sold off or neglected and the tasks taken over by contractors. Now, finding the work unprofitable, contractors are disappearing and farmers are trying to repair their old machines, as the cost of replacement is immense.

The biodynamic preparation known as 500 requires yet another machine, one that's not sold by local agents. You can't wander around a dealer's yard and kick the tyres. Biodynamic farmers have long made their own, and occasionally supplement their modest income by making some for sale.

Because I'm not mechanical, I have a deep appreciation of those who are. While I can grasp the essentials of what the eccentric-looking equipment does, I can't fathom the complexities when it inevitably falters. By 2004, with our olive enterprise doubling in size and the beef herd growing again after the drought, we had to face the fact that we needed new and bigger gear, particularly to spread the 500 further and wider, over more

500 is stored in a copper that sits inside a double wooden box lined with peat moss to ensure it stays cool and fresh

CLOCKWISE FROM TOP LEFT 1 Gavin getting the temperature right in the stirrer the old-fashioned way, by pouring in hot water with a bucket. 2 Lighting the new system. Heated by gas, it's much safer and quicker. 3 Checking the water level. 4 & 5 The stirrer creates a vortex, in response to which it changes direction, creating chaos. The whole process lasts an hour

distant paddocks and hills. And before we could spread it, we had to add it to warm water and mix it in giant stirrers.

I enlisted the help of other biodynamic farmers to build the gear and install it. One built us a hot-water system to more efficiently heat the brew (when fired up it thunders like a rocket about to blast into orbit), while another positioned a massive metal stand on firmer, stronger foundations. Atop this we perched two giant stirrers so we could gravity-feed the 500 into our mobile spray units.

Soon we were seeing evidence of increased health in the grasses and in the olive grove. Biodynamic farming creates healthier, softer, sweeter-smelling soils, ready to give rise to life in the way that a good risen dough gives life to bread. There comes a time when, walking over the paddocks, you can feel the land springing back, telling you it's alive. When I walk through the olive grove and feel the soil yielding to my steps I know the roots of the trees are digging deeper. It's good to sink a spade into the soil and see evidence of health in worms, bugs and beetles. Of course, a dry spell will bake it hard as a biscuit.

Truly and proudly a cottage industry, biodynamics is about releasing the natural energies, micro-organisms and nutrients in the soil, rather than forever adding artificial fertilisers. The core ingredient of 500 is cow manure, but a long process of maturation greatly intensifies its properties, and its chief benefit is encouraging the formation of humus.

Gavin adjusts the low-pressure spray unit

Superphosphated farms, with their brightly coloured paddocks, look as unhealthily pumped up as athletes on steroids, whereas biodynamics is slow, slow, slow. It's even slower than the Slow Food Movement, which biodynamics precedes by seventy years.

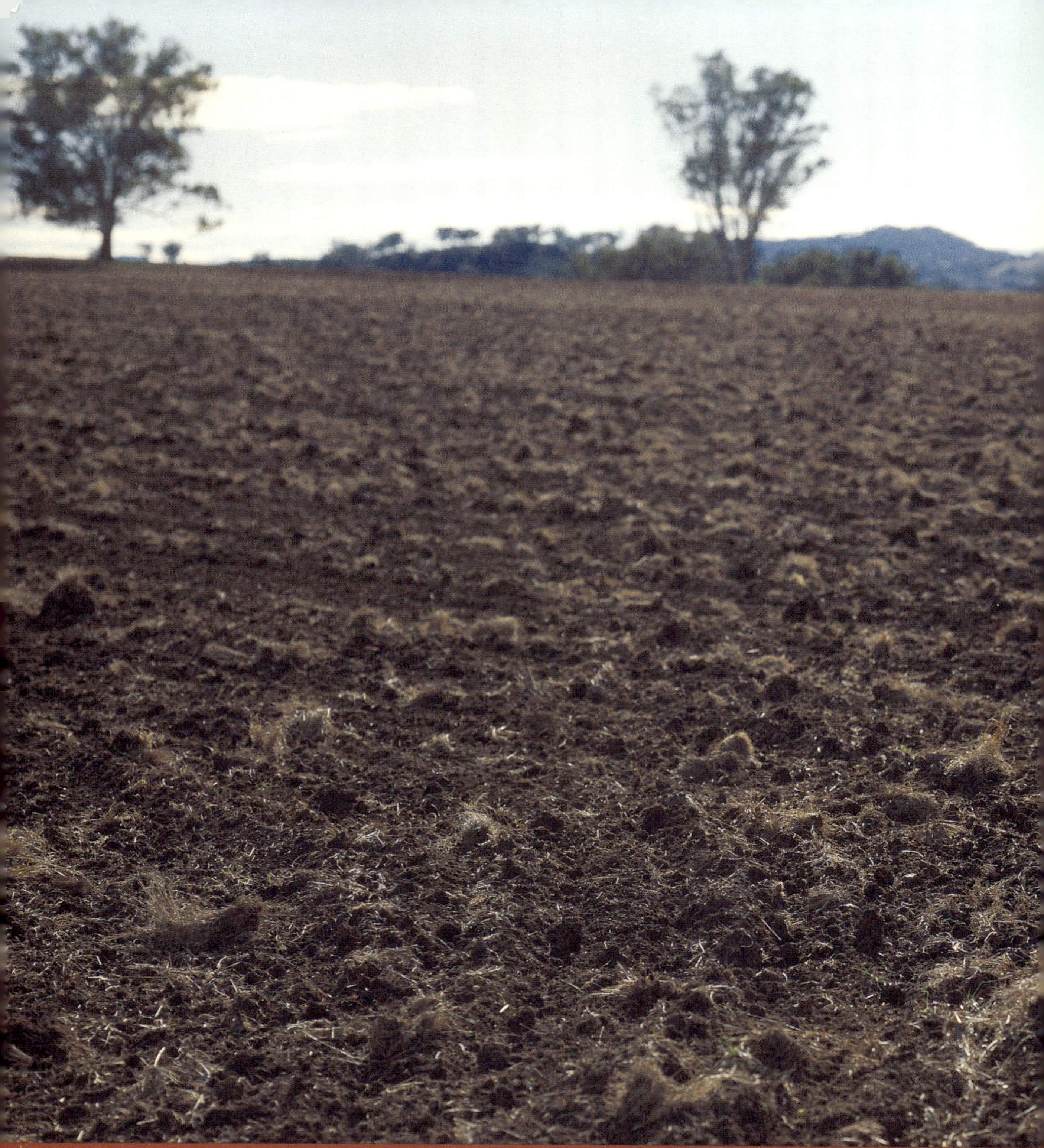

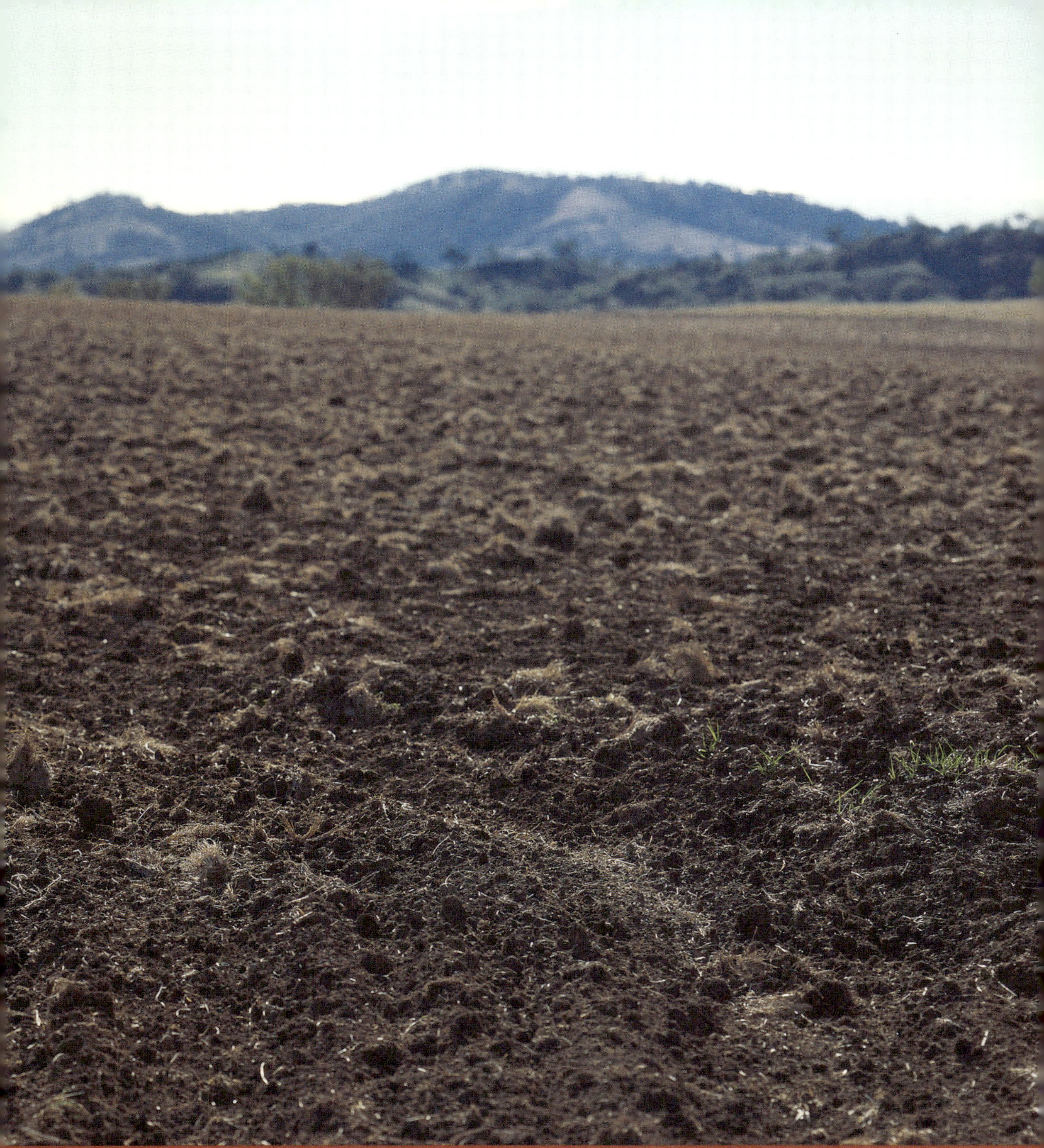

A mega-sized meat company invites producers to a gabfest to discuss the beef industry in the era of mad cow disease, free trade, the Atkins diet and a strong Australian dollar. I find myself with a hundred others in the alien setting of an RSL club. With pokies pinging in the background, we're told that much of the company's profits come from 'grinding' for McDonald's — that is, making mince. And there's been a surge in sales thanks to the hunger of soldiers in Iraq. 'Roll on the war,' the executive says.

The company buys young grass-fed cattle and herds them into feedlots, where they're force-fed a diet of grain. One breeder wants to know why the company won't buy his young grain-fed cattle.

'If they've been fed grain when they're young they just can't take it in the feedlots,' explains the executive. 'It strains their organs. Their livers give up.'

What am I hearing? An admission that they're feeding cattle food that kills them? At lunch we're served grain-fed beef but I've lost my appetite.

Ten Thousand Acres
A Love Story

- Ironbark
- 9/9/01
- side road at Ridge Rd.

On exactly the same day as last year, two frogmouths arrive and take their place side by side on their favourite branch in their preferred ironbark tree. I look up at them. They look down at me.

A few weeks later there are four of them. Two big-eyed babies. Things are as they should be.

Podargus strigoides
tawny frogmouth

I'm driving a friend home at dusk through the back of our property. The sun is setting behind us, the sky ahead is turning pink, and we're following the track along Kewell Creek, through thick stands of she-oaks with hills pressing in. My friend, who spent years filming wars around the world, says, 'I hid in valleys just like this in Bosnia.'

But this land tells no stories of local war. It's not like Europe, where every other piece of the landscape has been a battlefield. Even on Anzac Day, the war stories told in Gundy are about distant lands. What I've been unable to discover is whether violence occurred here at the time of white settlement, or whether the Wonnarua people retreated into the hills and kept retreating.

We don't keep cats to control rats and mice, not when feral felines are such a problem. That's another reason for being so fond of owls.

Over the years we've had plagues of mice verging on the biblical. Thousands of them eating oats in the silo, nesting among the hay bales, leaving their droppings in the library. We've set traps, which have a habit of going off while we're reading quietly, sending us jumping out of our seats.

Our first owl was a dead one. I found it in front of a wooden bench in the garden. From its spotty markings it was easy to identify as a boobook, but there was no indication of how it died — no blood-stained feathers or the like. Since then we've averaged about one dead owl a year, always finding the corpses in the same vicinity, but there have been other kinds: a barn owl, a masked owl.

We also have owls living on our front verandah, uniquely camouflaged by perching on a Sepik River carving. Collected by Phillip forty years ago, it has a shaft in the shape of a crocodile reaching up to grasp a winged human in its jaws. Convinced of their invisibility, the owls allow me to get quite close. Later, when I go outside to turn off a sprinkler, I feel them swoop close to my face. A slow wave of whooshing wings.

It's Travis who remembers reading about how owls have helped orchardists with their rodent problems. He produces a design for a nesting box and suggests we put one in the olive grove. Gavin hunts though our stacks of seasoned timber, pleased to find a use for the shorter ends, and slowly but surely the box emerges from a cloud of shavings — a thing of comfort if not of beauty.

We decide against the olive grove, opting instead for the shadows and shelter of the balcony — a place safe from cats, which we suspect, albeit without any real evidence, of killing owls. But it's not safe from the wretched currawongs, which we've seen attacking the white-faced masked owls. Aurora once saw an owl under attack pressing its heart-shaped face against her window, as if trying to get in. And we've heard them hissing as they swoop to escape.

So far, no owls have nested in the box but there have been plenty of visits, as evidenced by a discovery of furry white guano on the balcony floor.

I can't say whether the owls are eating any mice. I wish they ate kittens, because despite the best efforts of our dogs, the feral cats are multiplying. There's a thin grey tabby that perches in the cleft of a tree near the house; it looks a bit like an owl itself and enrages the dogs. That single cat is capable of killing hundreds of birds, and day by day it gets cheekier.

Early one morning I'm on the phone to a friend when the tabby saunters out of the laundry, looks at me with a sneer and goes arrogantly on its way. I curse it.

'Perhaps it's the neighbour's,' my friend suggests. 'Why not try and make friends with it?'

'No way. It's one Phillip's been trying to shoot for years.'

'I like cats,' my friend says. 'They're better than men. Soft and cuddly and they don't fart in bed.'

Contrary to cat lore, the ferals don't do much to keep mice numbers down. It's the wrens, finches, silver-eyes, and willy-wagtails they prefer. And, I suspect, owls.

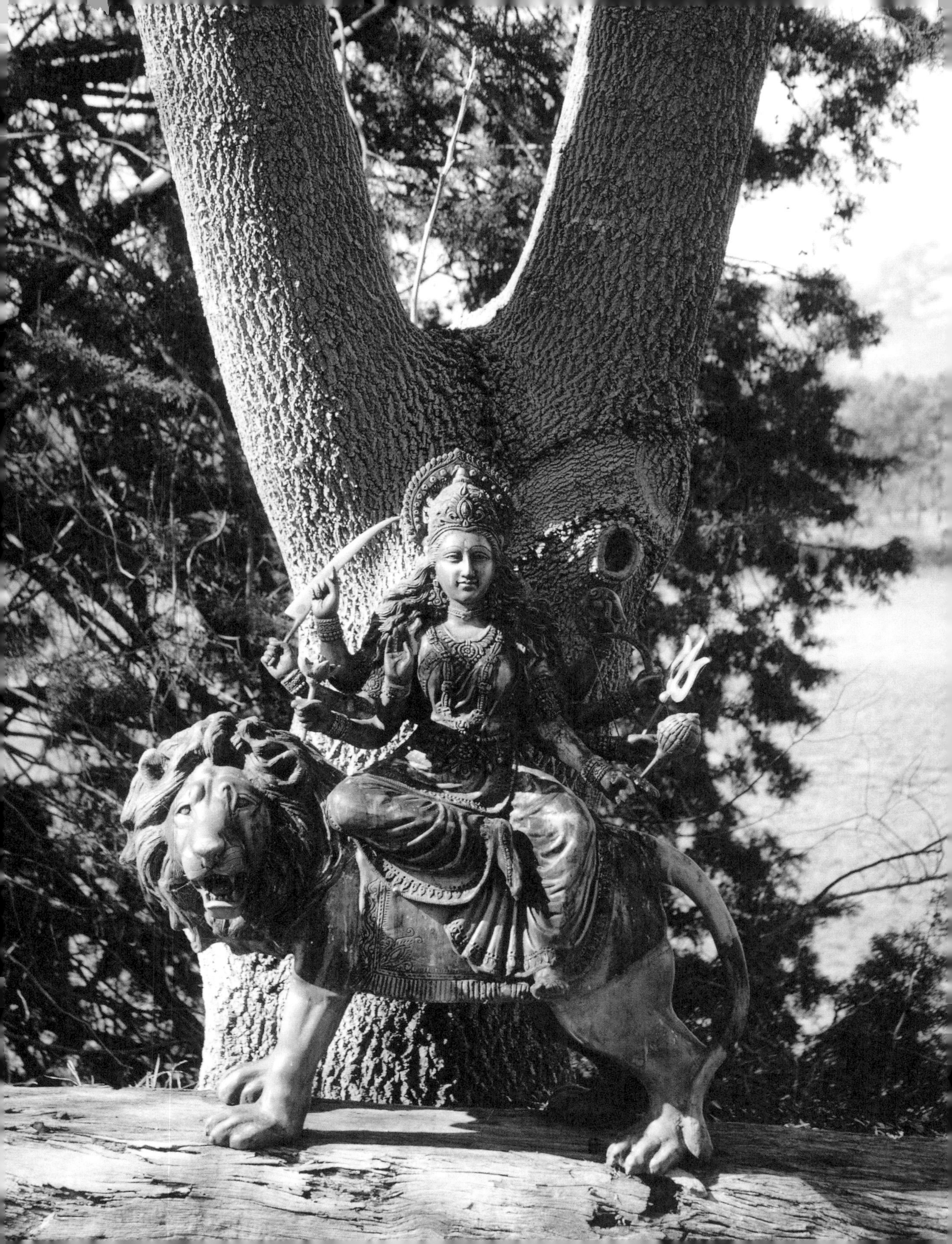

Assailed by doubt, I reluctantly agree to have a film crew from the ABC's *Australian Story* follow me around for a few days. They're interested in why I left the world of television for life on the land, and want to film me walking in my special places. I suggest a visit to what we call the Valley of the Dinosaurs, where steep hillsides are crowded with grasstrees – the ancient *Xanthorrhoea*. Countless thousands of them, surreal and spectacular, mingle with a wattle, *Acacia decora*, as bright as a cockatoo's sulphur crest. It's the perfect time to see both flowering, and the air will be soft with scent and thrumming with bees.

Phillip drives the crew up the perilous, rock-strewn tracks. They unpack their gear and I walk the cinematic walk, squeezing between black trunks on narrow paths created by kangaroos, trying to look as if I'm utterly alone.

Then it's time to record the 'atmos'. This involves standing perfectly still while the sound engineer tapes the silence, the particular silence of this particular place and time. Every silence is different, every place has its sonic fingerprint that editors need in post-production. So for a few minutes we all become statues and the engineer points his microphone, a thick sausage covered in grey fabric, down into the valley.

And there, a few feet away, just below us in the canopy of a eucalyptus, is the matching grey fur of a koala – the only live koala we've ever seen on the property. Delighted, the cameraman films it, but when the program screens the locals are suspicious. Where did we get it? they ask. Was it stuffed? Did we nail it to the tree?

Phillip takes Aurora back to the valley the week after the shoot, convinced that if there's one there have to be lots. But the branches are empty. We've been looking ever since, but apart from once hearing a distant roar that might have been a sexually aroused male, and the corpse that we mistook for a quoll, we've been unsuccessful.

We're warned by locals not to admit to having koalas, for fear of land grabs by officialdom, but I want to shout it from the treetops. From every koala tree, every *Eucalyptus tereticornis*, *E. albens* and *E. noblis*. Travis dismisses my enthusiasm for koalas as 'species engagement', by which he means I've fallen for the soft and cuddly. He's more interested in an unfamiliar beetle he's come across.

Xanthorrhoea glauca subsp. *glauca* grass tree

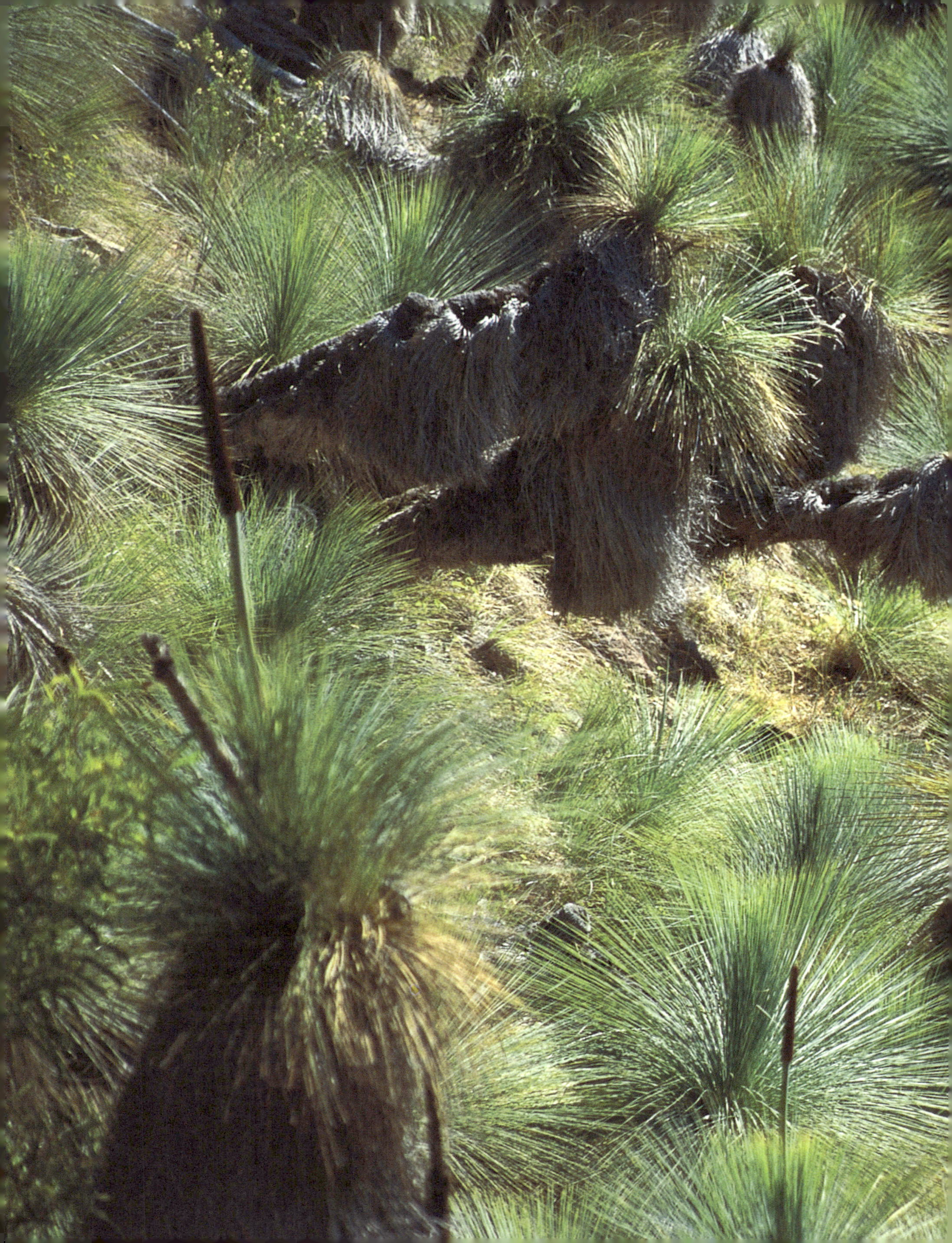

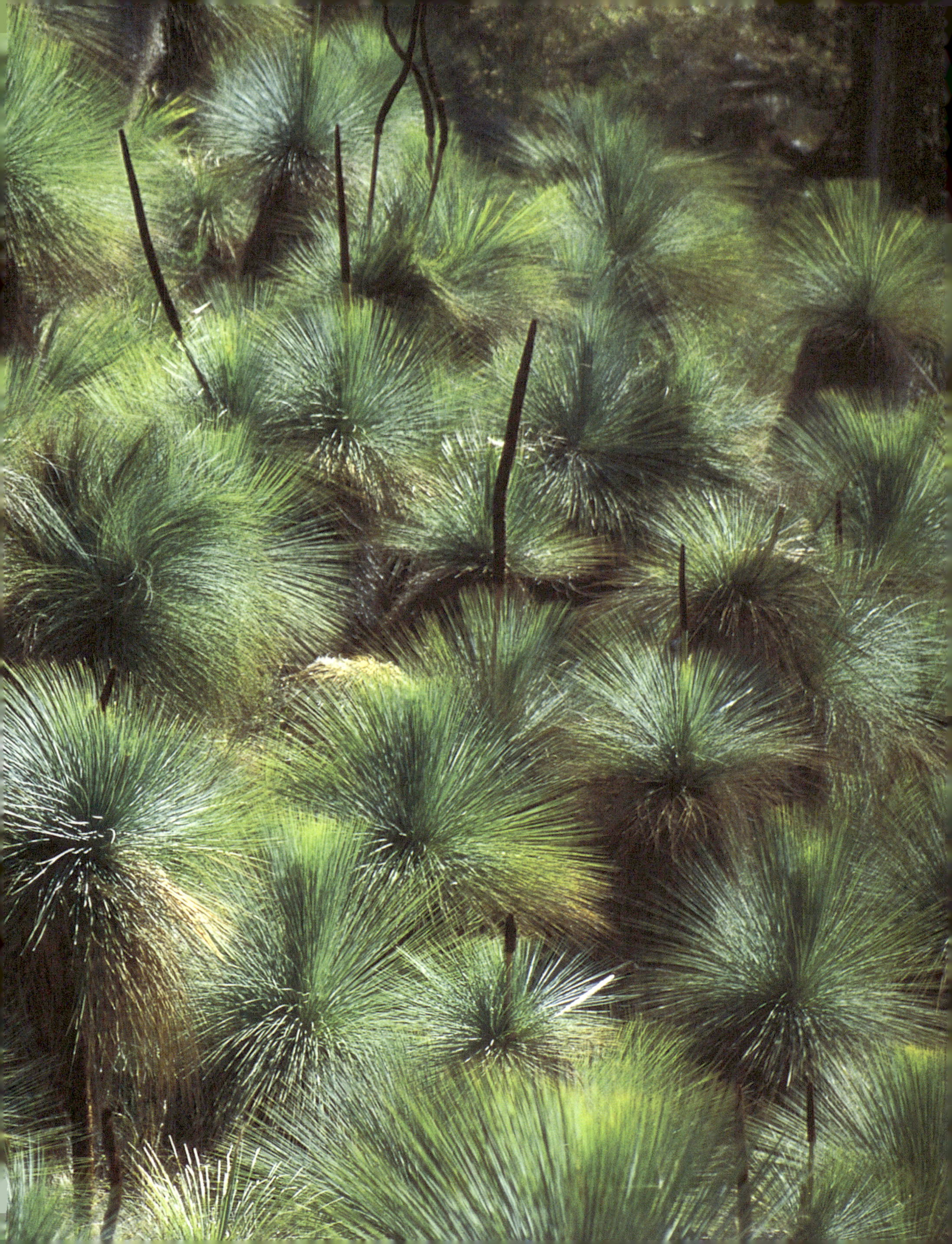

We bought horses soon after we arrived at Elmswood, learning to walk them up and over hills, to wend our way along dry creek beds. Phillip was self-assuredly carefree while I rode cautiously behind. Years on, I didn't feel I'd improved much; I was still riding at rocking-horse level. But at least I was getting more confident — until one windy day when Flint decided to buck me off, dumping me on particularly hard ground.

On my way down I remembered an ice-skating fall and another skater yelling, 'Tuck your chin in!' So I landed with my chin tucked in but still flat out, abandoned, spreadeagled, humiliated, emptied of air. A bruise in the grass. Then I remembered another voice from the past. 'Get straight back on the horse. Show it you're not scared.' But I was — very. I climbed tentatively back on and rode nervously, painfully home, all affection for Flint gone.

A few weeks later, I was back in the saddle at exactly the same place in the same paddock and Flint began to buck again. I slid off and walked him home, each of us giving the other looks of loathing. Warning the new owner of his personality flaws, I sold him immediately.

Months passed and I got a new horse, Tequila. When I put my boot in her stirrup for the first time I was shocked; I'd forgotten how high a horse was. I climbed back down and stifled a sob. It seemed that horses had become a symbol for everything that scared me, and I wondered if I'd ever overcome my fear.

Yet Tequila and I shared an affection for each other. I had lessons, and rose to the occasion as if I'd never ridden before. With Tequila's friendship I learned to trot, canter, feel my balance, give clear commands. I even learned how to stop.

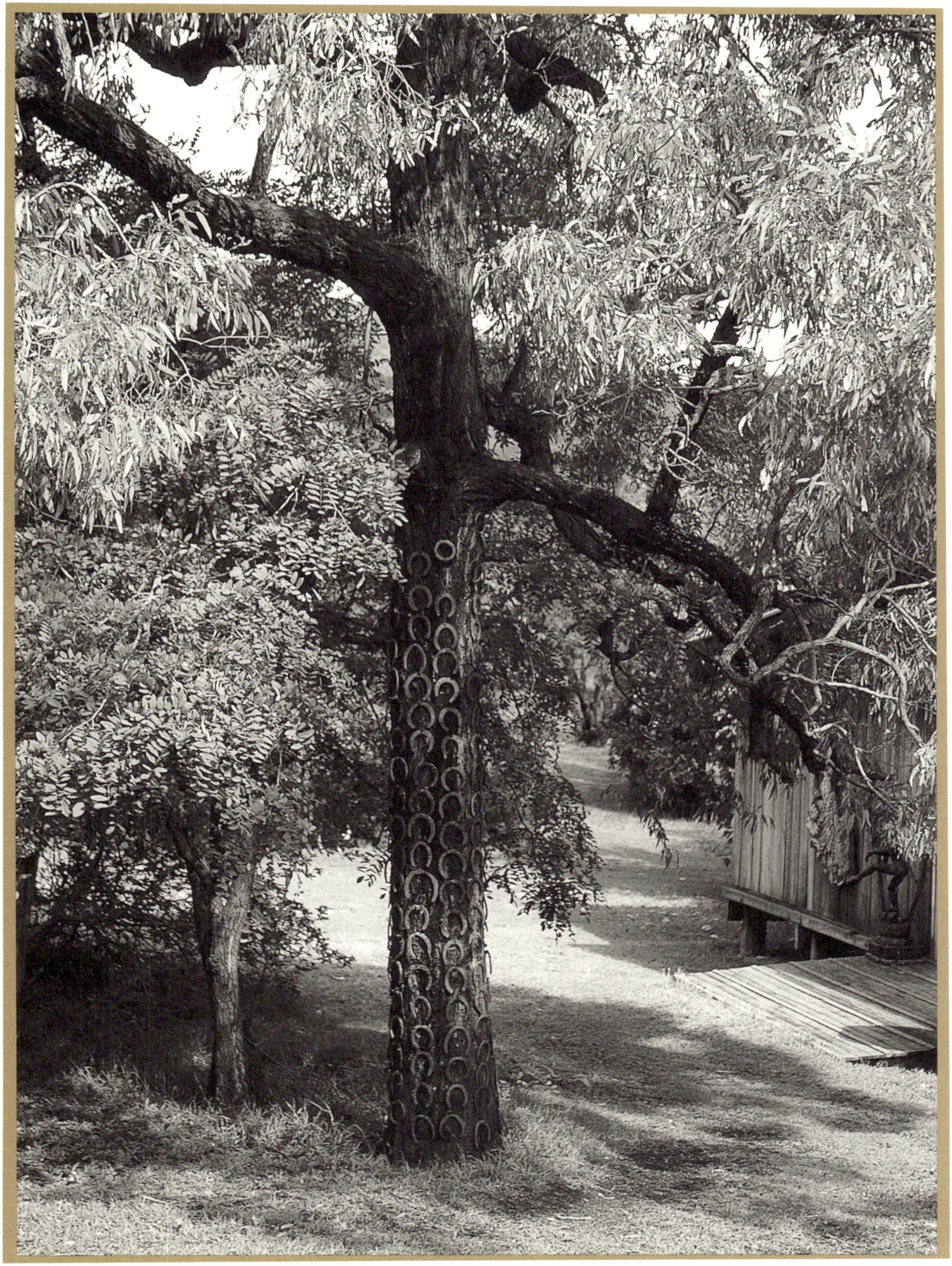

The dogs of Elmswood.
CLOCKWISE FROM TOP LEFT THIS
PAGE 1 Tommy, exhausted.
2 Chasing a kangaroo in the
olive grove. 3 Aurora pats
Tommy while Phillip cradles
George. 4 George a few
days before he was killed by
a kangaroo. OPPOSITE Rosie

Hearing the dogs barking frenziedly, I went to investigate and discovered them circling something by the back door. A snake? A blue-tongue? An echidna? No — a giant yabby, twice the size of those I used to catch in Adelaide as a kid.

Waved greetings were coming from a rusting bucket in the yard — claws wildly semaphoring as yabbies struggled to be free. And the whopper that was holding the dogs' attention was not alone; large escapees were marching over the lawn. We discovered later that some Gundy kids who'd been yabbying in one of our dams had left them as a gift. They were too shy to knock at the door, but under cross-examination reluctantly revealed which dam was so abundant. They had secretly stocked it with yabbies from home, and each year, strictly observing principles of sustainability, had managed to harvest enough for a feast.

During summer evenings we tried our own luck with yabby nets — loops of wire weighed down with mesh, a piece of liver tied to each centre, and ten-metre ropes. Circling the dam, we tossed them, like immense frisbees, as far as we could. Mostly they crash-landed just metres away, but other launchings were so successful they dragged the rope with them, sending Aurora leaping fearlessly into the ooze to retrieve them. But no matter how often we tried, we were never as successful as the local boys. As with the initial stocking of the dam, they kept their technique secret.

When I'm tired I close my burning eyes and listen, becoming attuned to the sounds, the music, of the landscape. Just as water has a thousand tastes, silence has many variations. Sometimes I hear owls sitting motionless in the trees, although they're not even ruffling their feathers. But mostly the air is full of the pizzicato of parrots, the adagio of insects, the soft strumming of leaves. There's something soothing about listening; it's less demanding than looking. Sound is as spatial and three-dimensional as vision. The close-up call of a bird contrasts with the wind moving in from distant hills. Sound is all around and inside you.

Himantopus himantopus
black-winged stilt

There's another report to add to the pile on my desk. Yet more consultants advising the government yet again that farming is in decline. It's time to move on, claims the anonymous author of this latest report, to abandon centuries of tradition. Forget the sentimentality, scrap the romanticism. Traditional farming is a fantasy. Get over it, says the report.

'Is this an economic truth?' I ask Phillip.

'There's no such thing as an economic truth,' he says.

Ominously we learn that the local abattoir will no longer kill sheep. This is for bureaucratic reasons: changing standards must be met and the abattoir cannot afford, or does not want, to meet them.

Farming may be in decline but bureaucracy is booming.

The wind is knocking the wind out of me. I'm in a boxing match, losing every round. I am a featherweight, the wind is a heavyweight.

When I try to hoe my vegie plot the wind whips dirt into my eyes and mouth. Birds have surrendered the sky, bees are in their hives, the cattle huddle. Branches are flailing and falling and the olive trees are showing their silvery underleaves as they bend and bow. Doors are slamming all over the house and I struggle to haul up the awnings before they're torn like spinnakers. Phillip chases his newspaper across the lawn and the chooks struggle back towards their pen.

It's been like this for days. The rainfall has been reasonable but the wind robs us of the benefit. I feel like Van Gogh in the mistral.

I'm always careful to slam the door when I exit the truck, ever since Gavin left it ajar and a brown snake slithered over the tyre and straight inside to coil around the accelerator. When I drive to the water tank to check the level, the wind grabs the truck door from my fingers and almost rips it off. A thousand-dollar repair.

The bare branches of the European trees, and the thousands of bulbs — none of which I planted — are part of the history of Elmswood. That old idea I once had of planting a garden to match the era of the homestead is dead and buried, but I still love walking the pathways, especially in the afterglow of winter rain when the air is clean and the sky blue. The ground turns green and the lawn is transformed into a flowering meadow

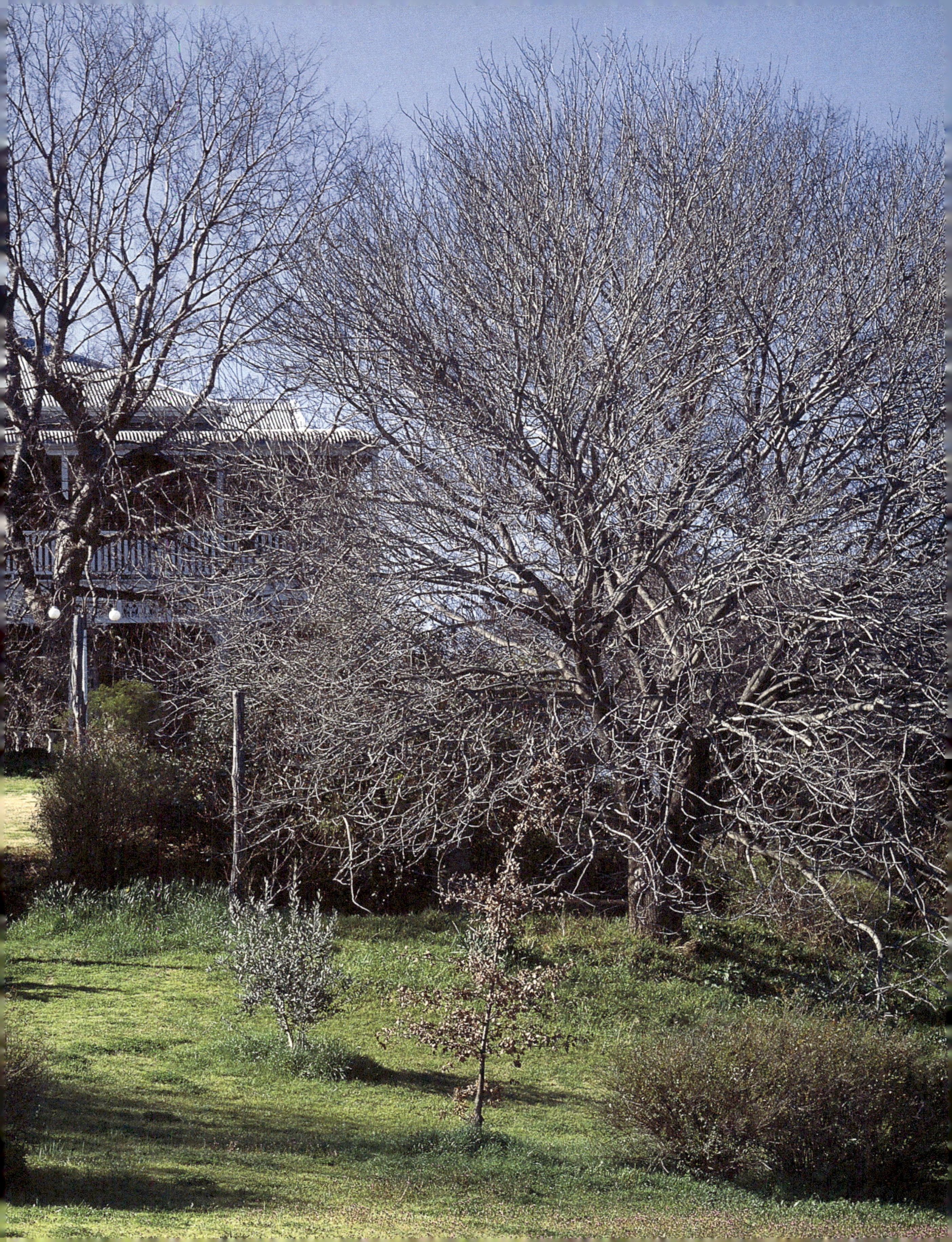

Eucalyptus melliodora
Station Paddock.
2/2/04

When I was a little girl I'd see one or two bee swarms every spring. Thousands of bees miraculously suspended by a thread in a backyard tree, humming, thrumming, threatening. My friends and I would play a game, daring each other to prod the swarm with a stick, or, even better, to swipe the swarm and make it chase us. Screaming with fear and excitement, we'd run as fast as we could, the bees — so we believed — pursuing us. I can remember being in a cloud of them, but strangely cannot recall being stung.

Yet there was the time I stood on a single bee while hanging clothes on the line and my foot swelled immediately. It wouldn't fit in a shoe for days, and it itched for weeks.

These days I treat honey bees with greater respect. We bought our first hive after a weekend beekeeping course, and for the past fifteen years we've been harvesting honey. Hundreds and hundreds of kilograms of molten gold. Despite a deepening fascination with the life of the hive, I cannot claim to know bees much better. I'm familiar with individual cattle in the herd, with their personality traits, but our bees remain anonymous, the hive enigmatic.

Most of our hives are in front of the house, near the olive grove, where I can keep an eye on the comings and goings, observe the degree of busyness and the quantity of pollen on the bees' legs. Every few weeks I fill the smoker with dry eucalyptus bark, get it smouldering, and huff and puff at the hives, carefully opening the lids and lifting the frames to see what's happening. Who needs Tolkien when you have a magical, mysterious beehive?

I've arranged the hives so that the bottom box is for breeding and the top for harvesting. The queen takes a few flights from home to copulate with some of the drones, killing them in the act and returning with the severed sexual organ of the last drone attached to her like a trophy. Her Majesty then spends the rest of her life laying fertile eggs, which, depending on the way they're fed, produce either workers or queens.

During summer the hive is a labour ward, full of bees being born. Having wriggled from their cells, they immediately set to work, taking wax from their stomachs and massaging it with their legs until it's soft enough to make new cells, the hexagons of the honeycomb.

Older bees shuttle to and from the hive delivering nectar, a watery, sugary solution whose character varies according to its floral source. Crouching by the hive, watching the bees through my visor, I see the old bees kissing the new bees – but actually they're transferring nectar. The young ones then spit it into their wax cells, where enzymes from their salivary glands thicken it while they energetically flap their wings to change the moisture level. When the right consistency is achieved you've got honey.

Bees are endlessly foraging, guarding, cleaning their home and each other. A single bee can visit a thousand flowers a day. Having made a discovery, it will do a little dance to show the others where it's been. Remarkably, during a lifetime of five or six weeks, involving visits to as many as forty thousand flowers, one bee may produce just a single teaspoon of honey. So every mouthful should be savoured.

Honey is mostly sugar, water and hard work, with traces of minerals, vitamins, proteins, acids, enzymes, alcohols and esters. Volatile aromas give honey the smell of the flowers of their origin.

We have a yellow box, *Eucalyptus melliodora*, by one of our sheds and when it flowers, which it does erratically, it literally rains honey, showering us with scent. Beekeepers talk of times like this, when plants fill with nectar, as a 'honey flow'. You can see the bees getting busier, their movements quickening as word of the nectar spreads. All is urgency, and the harder they work, the happier we humans are. We admit to robbing their hives. And I for one always feel a little guilty.

When the boxes on top of the brood box fill with honey, Gavin, Aurora and I arrive in our space suits, brandishing our smokers like priests with incense burners. We have to learn the lesson of any sophisticated parasite: not to rob too much. We have to assess the volume of the honey flow, its likely duration, and consider the weather and the subtleties of seasonality. The most important factor is the 'strength' of the hive. Is the queen laying eggs vigorously? Is the population ageing? Are there enough young bees? You deliberate, you calculate, you guess. It's imperative to leave enough honey for the hive itself, so that the community can cope with cold snaps or dry times. Should a beekeeper rob too much honey, or the nectar supply falter, they will often feed their bees sugar to keep them going. We never do.

Honey is so heavy I cannot budge a whole boxful of it. I have to remove just a few frames at a time, brushing away the bees so they won't follow me home. Standing by the extractor in the shed, I hold the frames firmly on their sides as Phillip scrapes loose the capping. Then he fits each frame into the extractor and turns the handle. The thinnest of filaments, finer than any cobweb, forms a glistening skein of honey on the walls. On and on the spinning goes, and we smile at each other as the extractor fills.

This is just about our favourite job together on the farm. Each extraction produces a slightly different honey, reflecting the nectar and the season. Yellow-box honey has a low glycaemic index and is recommended for diabetics, while that made from *Angophora floribunda*, rough-barked apple, is darker and less complex. Sometimes the honey is runny, simply pouring out, but during droughts it's thicker and more intense.

Environmental journals report that European bees steal tree hollows from native birds and marsupials, and argue for poisoning their feral swarms. I've seen escapees buzzing in tree hollows and wondered if they once lived in my hives. Like cattle, like humans, they're now part of our landscape. Are these bees destroying the pollination of the native flora around here? I simply don't know.

When cooking I'll often replace the sugar in a recipe with honey, reducing the liquid by a quarter of a cup. Sometimes we use honey instead of antiseptic cream, and it's common for farmers to make honey poultices for injured horses. According to folk medicine, if you eat the honey from your area you'll slowly absorb the pollens that cause allergies and build up a resistance to them. When we first came to the farm Phillip would almost asphyxiate from hayfever, but after we began our own honey extraction and he ate a lot of it on the comb, his hayfever disappeared.

Nothing our bees produce is wasted. We save every scrap of old wax, melt it down and produce perfect tapered candles. Renoir was right: the light from a beeswax candle is far lovelier than the almost fluorescent glow you get from tallow.

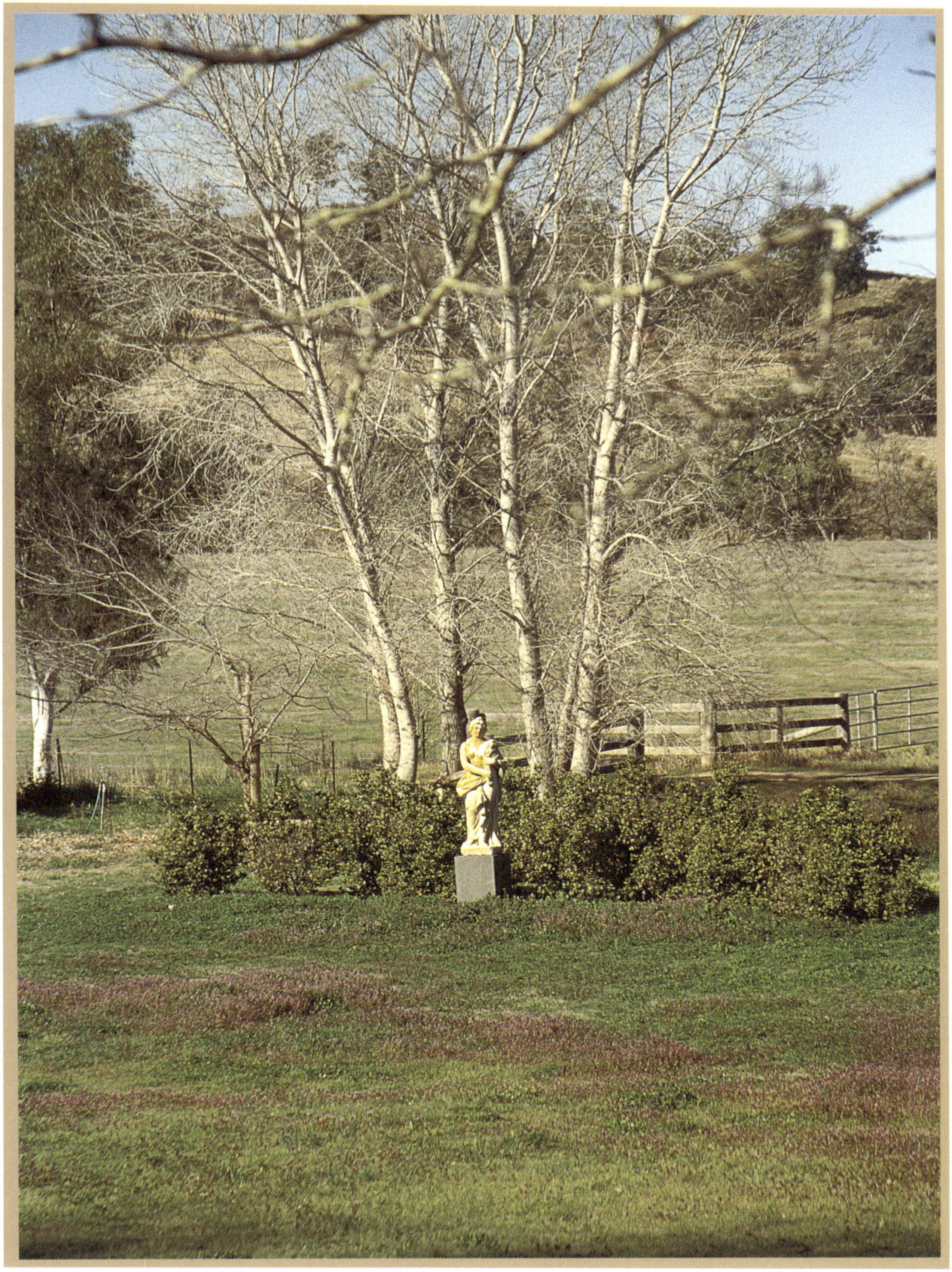

When my mother died I found a heavy cardboard box under her bed with 'Aurora' written on it. Inside was something wrapped in a towel, and inside that was a parcel in butcher's paper with 'tomahawk' scrawled across it. Aurora and I looked at each other and wondered. Then curious fingers got to work and — a little axe. Very old, very sharp, with a worn handle.

I half remembered it. Had it belonged to my grandfather? It was an odd gift for a seven-year-old, but Aurora was enthralled.

Mum preserved all her father's old tools lovingly, carefully. Previous generations relied on their tools, respected them. These days they're so cheap they're as disposable as Bic razors. Take trowels — I have half a dozen in the vegie garden and use them as markers to remind me where I stopped work last time. At best the oldest linger on as garden sculptures. I rarely wipe off the dirt and return them to the shed, as Mum did, oiling their wooden handles and hanging them neatly on nails.

There's only one tool I treasure — the secateurs Phillip bought me twenty years ago. We hand out secateurs during pruning and many don't return, but I keep my Felco pair separate. Beautifully designed (the handles oscillate in the palm to prevent blisters), they can be kept sharp for ever with a small file, and they've cut just about every plant I've ever grown. One of these days a grandchild will inherit them, wrapped in a towel in a cardboard box.

Few people have a kind thing to say about our native bracken. The word itself sounds harsh and farmers see it as the enemy. It can cause a Vitamin B1 deficiency in cattle and can also produce haemorrhaging. 'Those pigs are hiding in the bracken,' farmers will say. Or, 'We're losing our hills to bracken.' And if you take it on bracken fights you to the death. You can slash it and slash it in the hope of creating a beautiful hillside of native pasture, and the bracken will always come back. But I have a friend in Tasmania who uses a twist of the wrist to tug bracken out all through his garden — to ensure he's damaged the runner — and he's succeeded in controlling it.

Yet to look at a single plant of *Pteridium esculentum* is to marvel at its delicacy. As on a tree fern, tiny fronds unfold like a shepherd's crook.

Other ferns grow at Elmswood — wisps of maidenhair, *Adiantum aethiopicum*; and poison rock fern, *Cheilanthes sieberi* subsp. *sieberi* — in shaded crevices and on clear slopes in full sun, but bracken is the most assertive. Ferns are the genesis plants, survivors from Eden and the dinosaur era. Every plant since the fern is a new arrival. Certainly all the flowers.

Pellaea falcata
sickle fern

Adiantum aethiopicum
common maidenhair

Forty years back, Phillip had some villagers in Ubud make him a small army of huge temple guards — elaborate, fearsome monkey gods, all claws and fangs. Just as it took the Balinese weeks to drag the blocks of sandstone up the hills for carving, it took us weeks to position them around a pathway in the garden. Standing back, considering, Phillip decided to connect a tiny drip system to their heads. The idea was to trickle water over them in the hope of encouraging moss and lichen. I would take the odd bottle of spoilt milk or jar of old yoghurt and pour it over them to help the process.

While some delicate lichen did grow over their fearsome faces, the patination died every summer and we gave up. Instead I dug maidenhair fern from the hills, transplanting it in clumps around the statues' bases. The complexities of the carving proved a perfect home for plants that like crevices between rocks.

Maidenhair evokes the idea of moisture, yet I've never seen the fern wet. Somehow water beads up and rolls off, leaving the plant bone-dry. It nonetheless creates an illusion of tropical humidity that the monkey gods seem to enjoy.

Pellaea falcata
Sickle fern.
Feb 2004.
Kewell Creek

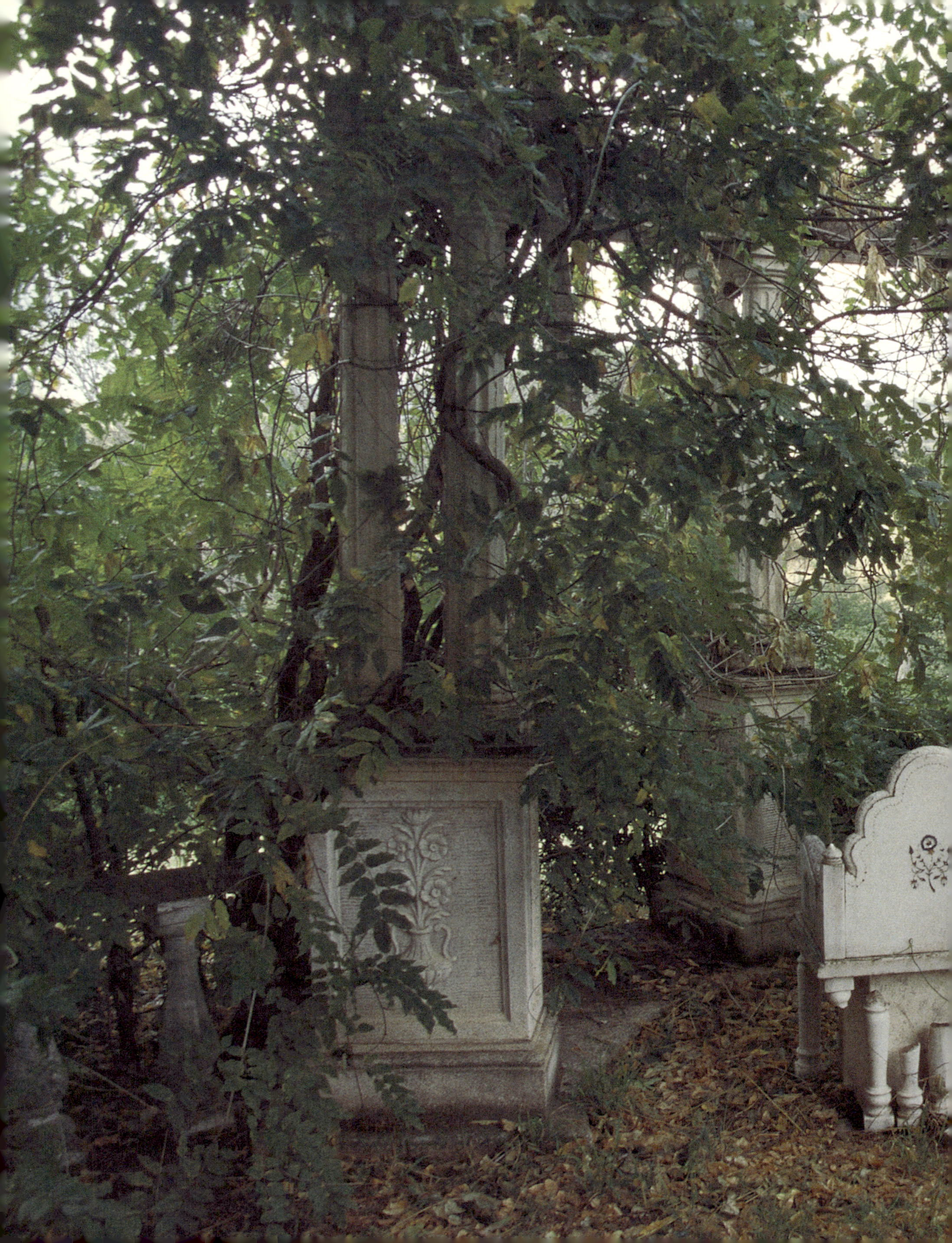

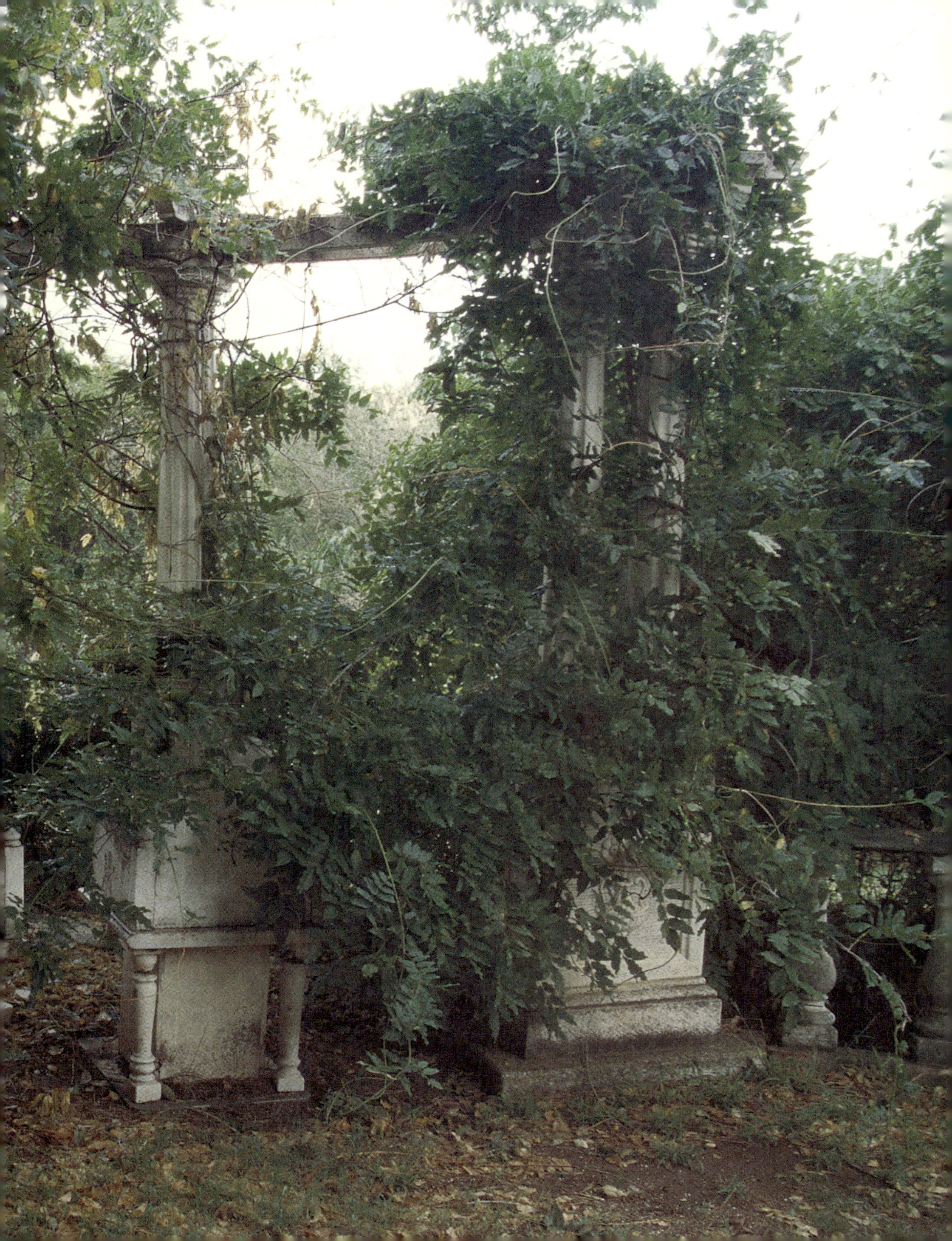

Travis parks his truck under a huge bough of a rough-barked apple, *Angophora floribunda*, about to burst into flower. It's little wonder people confuse this tree with a eucalyptus. It is, after all, a close relative. From a distance the bark, leaves, colour and flower buds evoke the gum. Up close you can tell which is which by the position of the leaves: on the rough-barked apple they're opposite each other, whereas the mature eucalyptus has alternating leaves. Angophoras don't have caps on their flowers, where the eucalyptus does, and they don't turn their leaves to the side as much, to hide from the sun. They simply won't fit nicely into a botanical box.

Because they begin to bloom in December — when the flowers, filling with nectar, can weigh the branches to the ground — I call them our Christmas tree.

Angophora floribunda
rough-barked apple

The narrow-leaved ironbarks, *Eucalyptus crebra*, have looked pregnant for months. Buds bursting with intention. Now half the hillsides are covered in their creamy blossoms and glossy leaves. Is the timing programmed into their genes?

One of the first plants given to me when we came here was an unlabelled reed that I planted beside Phillip's lily pond. Over the years, we've watched it multiply into a thick clump. During that time, Australia's municipalities discovered it and it became the official plant for roundabouts and freeway edges. As an approved native groundcover, it began to replace dietes and agapanthus. The scientific name of the reed is *Lomandra longifolia*, and we're still 'discovering' it all over the hills and along the riverbanks.

Every few months a witty newsletter arrives from Noel Jupp, who owns a nursery in East Gresford. Like Travis, he's been walking all over the Hunter Valley for years, collecting seeds and cuttings to ascertain their commercial suitability, and his nursery display attests to his eye for plants and the extent of his botanical interests. Noel writes about the ever-increasing varieties of *Lomandra* he comes across — evidence that this native plant is hybridising. He's invented a pseudoscientific term to describe these species-in-the-making: the CMU form, or crazy mixed-up form.

In nature there are boundaries that prevent genera crossing with each other — *Angophora* with *Eucalyptus* or *Corymbia,* for example. But species do intermingle, creating hybrids. When an area is disturbed by human activity, or by nature — a landslip or flood, for instance — hybrid swarms erupt, and suddenly new plants bob up. With the massive disturbances caused by mining in the Hunter Valley, there are lots of

Lomandra longifolia
spiny-headed mat-rush

new lomandras. Noel grows hundreds of thousands of plants for use in the rehabilitation of mine sites. Suitability is all important, as is getting the names for all the varieties right.

It's little wonder, really, that taxonomists are about as popular as tax collectors. They're equally bureaucratic, forever recategorising plants just when you've memorised their existing genus and species. There are two streams in taxonomy: splitting and lumping. Splitters celebrate the endless diversity while lumpers tend to group plants together. 'Are you a splitter or a lumper?' I ask Travis as he photographs an angophora. He thinks about it for a second. 'If I were a taxonomist I'd be a splitter.'

Angophora
floribunda
20 Dec 01

Records of early white explorers tell us that once upon a time, a hundred and twenty mighty trees stood where the Pages River meets the Isis. The Wonnarua tribe had turned these trees into an art gallery, covering their massive trunks with carvings. Now both the Wonnarua and the trees are long gone from this place. Only one uncarved *Eucalyptus tereticornis* survives. It's similar to the Murray River red, but not quite as majestic in size. It's perhaps three hundred years old, and broken branches have left it full of holes that provide homes for an assortment of birds, mainly parrots. Every flying visitor to this part of the river uses the veteran as a perch.

Fifteen years ago I planted a clump of eucalyptus nearby and already they're half its height. Spindly teenagers beside an adult. Whenever we plant trees on the farm we know we won't live to see them in their prime. Great trees need generations to witness them. When we stop for a picnic or a billy tea we choose a place near an old tree. They have gravitas, and perhaps it's their own force of gravity which draws us to them. Sometimes, driving around the farm, I'll see an old tree and book it for a future picnic.

Many of the oldest are *E. melliodora*. They're messy, dropping a huge amount of debris. Giant limbs fall in storms. As they age they look like wounded warriors, victims of crude amputations. The fallen boughs demand you sit on them, and when the bark sloughs off it reveals insect engravings of truly beautiful, elaborate design. I've planted three in the garden.

Mistletoe on a *Eucalyptus tereticornis*. Some ecologists call mistletoe a keystone plant because it's ecologically essential. The mistletoe bird, *Dicaeum hirundinaceum*, having ingested a mistletoe seed, sits sideways on a branch so that the sticky seed easily attaches. In some parts of Elmswood mistletoe is growing on trees that are only four metres high. Elsewhere it's in the uppermost branches of old giants — different species of mistletoe in different trees. All birds seem to love their nectar

Amyema miquelii
box mistletoe

Ten Thousand Acres
A Love Story

9/9/01
Acacia
Salicina
Middle Pd.

The spiders on our farm are huge. Could it be because we don't spray chemicals?

To prevent arachnophobia, I used to tell Aurora that we needed lots of spiders so that the willy-wagtails would have enough webs to make their nests. And it could be true: every shed is a willy's breeding site, where we watch them catch insects in the air. As a result Aurora is marvellously unconcerned about spiders, but people helping pick our olives are frequently aghast when their fingers touch an abdomen that looks uncannily like an olive.

I've grown quite fond of spiders, seeing our impressive population as a sign of environmental health. For me, the more the merrier.

With the best of intentions I'm always putting insect specimens in jars, but most lie forgotten in the backs of cupboards, where the labels fall off and the insects go mouldy. Clearly I'm better at preserving fruit than flies. But there is always another insect to capture – so many weird and wonderful things hopping, buzzing, flying, crawling, slithering and biting.

John Gollan, a student entomologist at the Australian Museum in Sydney, has been collecting insects in pit traps by the fairy martins' bridge. This is a rehabilitation site – our attempt to reintroduce native vegetation along two kilometres of the Pages River. John is comparing areas of different decrepitude to see how the insects are faring.

When Phillip and I first came here I had a simple test for determining the health of the soil. I'd sink the prong of a fork in and lift out some dirt, smell it, and after examining the depth of any roots present, the quantity of white feeder roots and the level of leaf mulch, I'd crumble it in my hand and count the insects. If there were at least ten different kinds, the soil was fine.

John puts my efforts with fork and jars and dead insects to shame. He leaves Elmswood with hundreds of examples to sort in the gloomy bowels of the museum, where he's surrounded by countless drawers of specimens containing numerous species. Preserved in alcohol, they wait to be pinpricked and positioned on trays before being identified and catalogued.

I drive to Sydney to see the ants, wasps, spiders and beetles that John collected at Elmswood. Today he's preparing beetles for the taxonomist, Chris Reid, to identify. If John thinks two individuals are the same species he puts one aside, but he has

The Pages River floods, rising two metres. Will I live to see a flooding that doesn't wash the soil away from the catchments? People talk about a flushing of the river as if it's therapeutic, but that brown torrent is like a loss of blood. At least the Pages has no traces of Prozac in it, as an environment agency reported of rivers in the UK. In Italian rivers it's cocaine

to be careful — make a mistake and he'll skew his species count.

Beetle identification and classification are not easy. Close, often microscopic, inspection of mouths or genitalia is required. I shiver in a cold back room that stinks of naphthalene as John slides open a drawer to show me yet more beetles.

I'm eager for answers to some questions. Was there any difference between what John found at Elmswood and what's been found at other riverine sites? What about the difference between the rehabilitated and the undisturbed sites, in terms of the quantity and type of insects present? Does it matter if some aren't present? If an insect isn't there, perhaps it's due to local adaptation, or seasonal movement? And can I have a list of all the insects he found?

Chris makes the point, calmly but tersely, that there are ten thousand insect species in New South Wales alone, and hundreds of them have not yet been named. 'It's more important to know what sort of insects they are,' he says. Leaning over the trays, he rattles them off. 'You have leaf eaters, grass-seed harvesters, wood borers, ant-nest inhabitants, casuarina feeders, sand burrowers, dung beetles — native and exotic. Plus predators, including snails, worms, caterpillars, cutworms and fly-maggot specialists. All pretty ordinary.'

They don't look ordinary to me. Some are as big as my thumbnail, others almost as small as the pinhead that's holding them in position.

Ozothamnus diosmifolius
white dogwood

Caladenia carnea
var. *carnea*
pink fairy

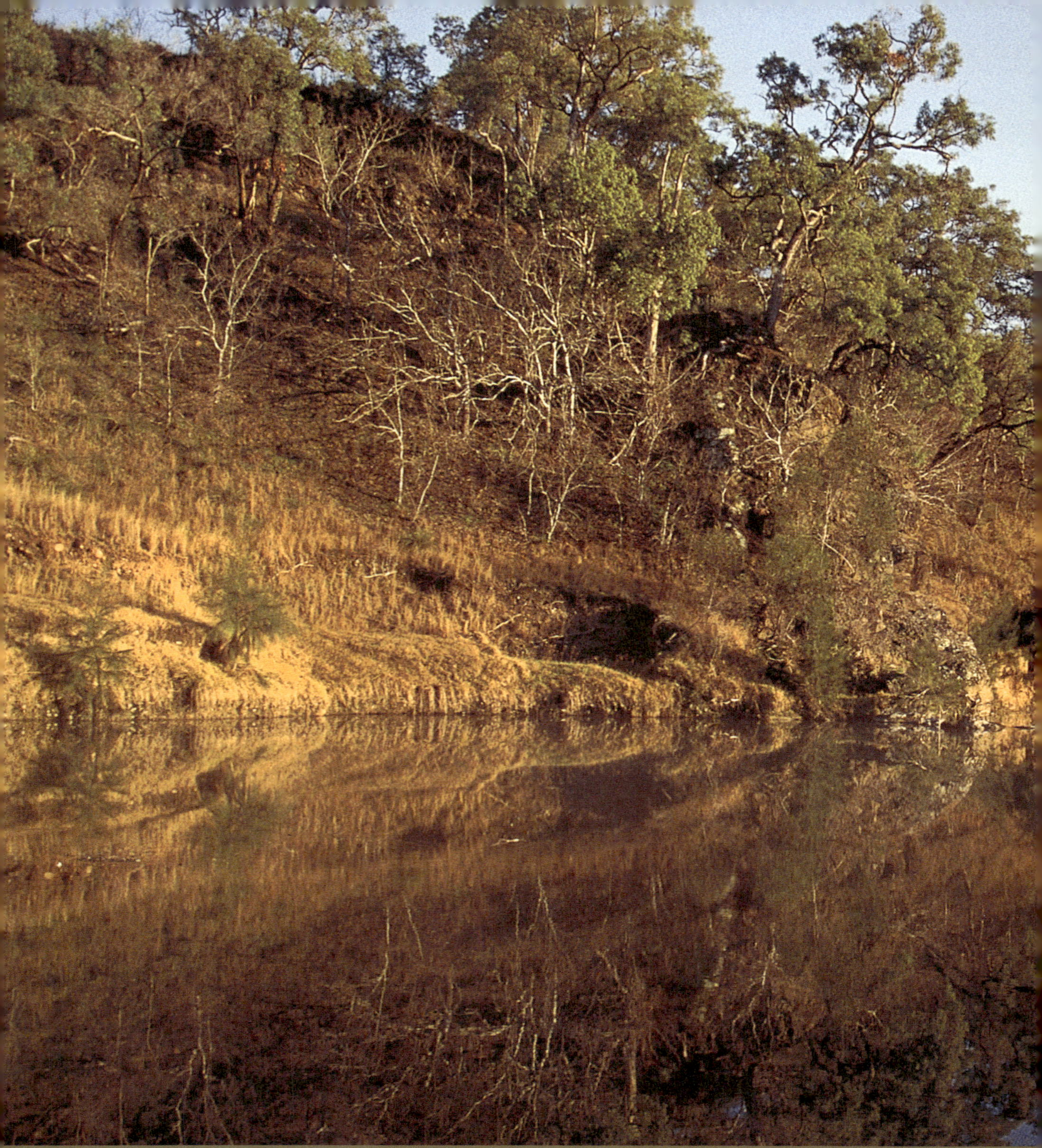

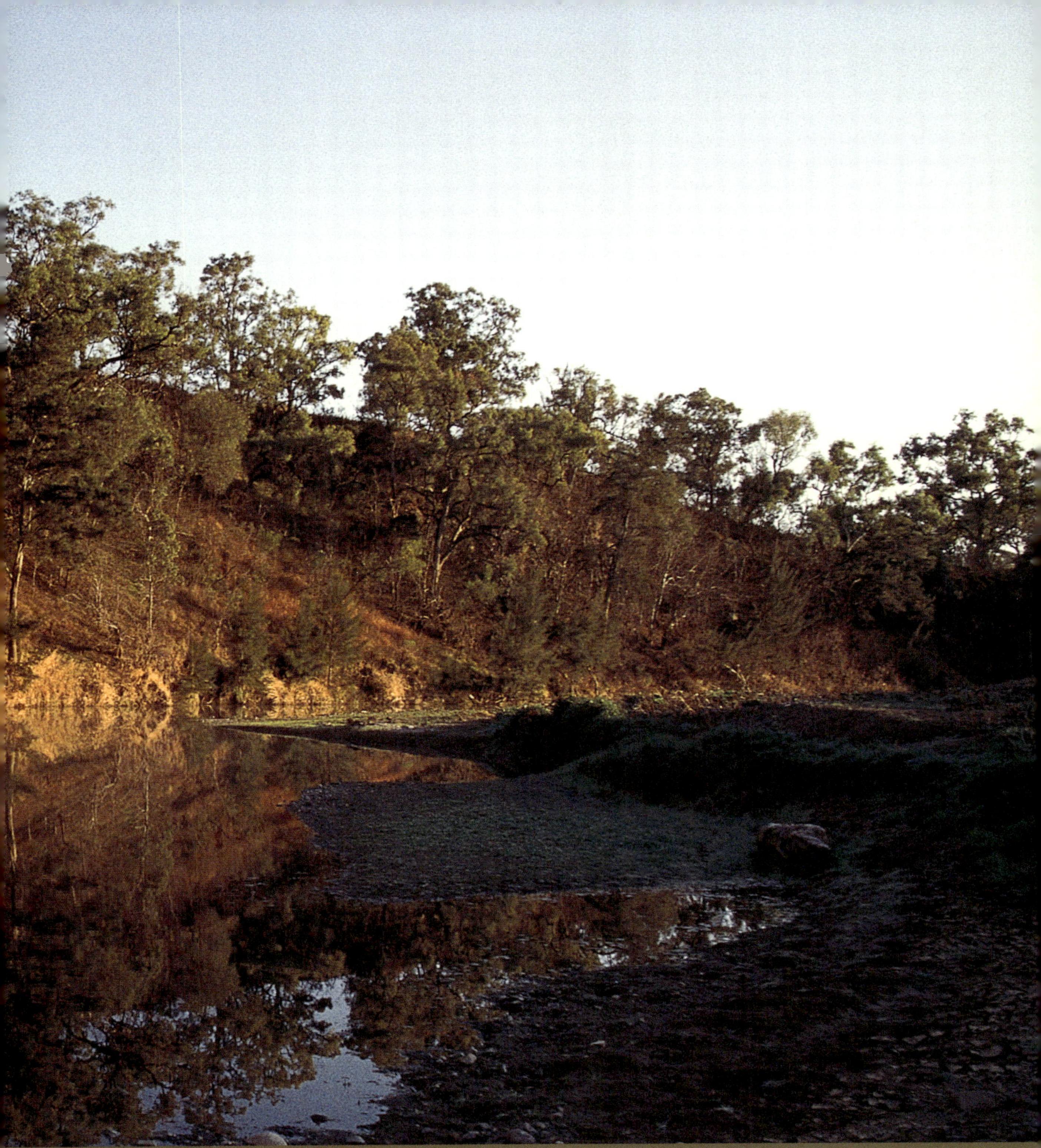

On some bare hills, dead trees are home to creepers that rise up their trunks like green fire. In the deepest gullies energetic tangles of vines and twiners would make Tarzan feel at home. Piles of boulders are wrapped in the writhing roots of fig trees, where creepers fight for supremacy.

If I took cuttings of every creeper, twiner, rambler and climber on the farm I could create a creepy, almost supernatural garden. There is something spectral and spooky about them all. So far I've succeeded in growing only two of them at home: *Hardenbergia violacea*, known as false sarsaparilla, and *Clematis aristata*, which, as I point out to dear old Phillip, is old man's beard. I'm trying to establish *Pandorea pandorana* subsp. *pandorana*, wonga wonga vine.

Our native grape, *Cayratia clematidea*, climbs over blackthorn and delicately clasps she-oaks. It ripens at precisely the time that the local vineyards bring in the harvesting machines. Some day, these small, tart, black native grapes will be a common bush food.

Hibbertia scandens
climbing Guinea flower

Clematis glycinoides
var. *glycinoides*
headache vine

Cayratia clematidea
slender grape

Hardenbergia violacea
false sarsaparilla

Pandorea pandorana
subsp. *pandorana*
wonga wonga vine

16/10/04. Spinvale P⊕

creeper big.

– cissus antarctica.
  water vine

Vitaceae

In summer it can be almost impossible to find a liftable cow pat. No sooner does one plop onto the ground than the dung beetles arrive, burrowing and perforating, devouring. The process accelerates with rain, and grasses sprout through the pats so they cannot be lifted. Within a few days they've returned to the earth.

When we plough during perfect conditions — not too dry, not too wet — it's like slicing a chocolate cake. The earth smells rich, delicious, edible. It looks as though you could scoop it up with your hands, arrange it on a plate and serve it with a pot of tea. And the earthworms agree. If I lift a cow pat with the toe of my boot, they're there working with the dung beetles.

Irrigation pipe being
laid for stage two of the
olive grove.

A friend arrives with food from her farm. A sack of freshly harvested wheat, new season's apricots, last season's marinated olives, a jar of dried rosehips from *Rosa rubiginosa* that we plucked last autumn from bushes growing on our hillsides. We once dreamed of turning this weed into a commercial product, making tea from rosehips. Like many an idea, it never ventured beyond the kitchen, but now I put the kettle on.

The wheat in the sack is biodynamic and doesn't have the musty smell of a grain silo or a sterile packet of flour from the supermarket. This wheat still holds the sunshine. I lift the electric stone grinder from the pantry shelf and begin converting cupfuls into flour. Turning the setting to extra fine, I grind some for muffins. I use a coarse setting for bread and pasta – this way you taste the wholeness of the grain.

I'll add our eggs, and the pasta will be perfect with a thick ragu of veal, beef or creamy mushroom. Or I'll toss in fresh herbs, squeeze on some lemon, and pour our olive oil over it.

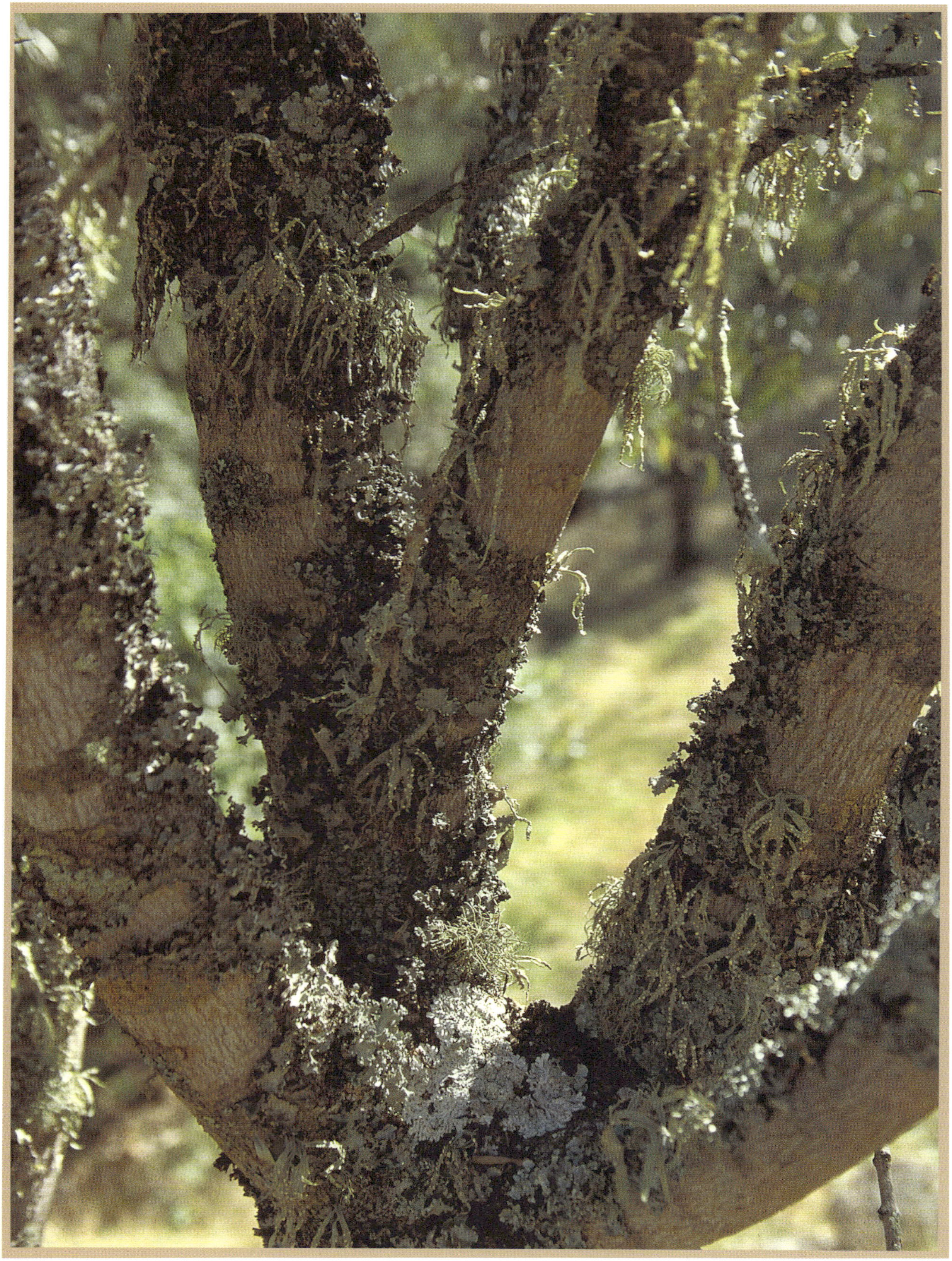

After autumn rains, we collect mushrooms in the sunshine. *Agaricus bisporus* are so evenly arranged on the hills, we might have sown the spores ourselves. We also find earth stars, giant puffballs, corals, brackets — and we take photos of other fungi, having no idea what they are.

Australia has the greatest diversity of fungi in the world, greater than that of our plants, although we fail to see the majority of them because they're microscopic — cobweb-like threads in the soil. Mycologists estimate there are as many as 250 000 species in Australia, most of them, like our insects, unnamed. Of these species, some five thousand are mushrooms, with an array of fascinating colours, textures and forms. They're strange, ephemeral things. And 'things' they are — neither plants nor animals, fungi occupy their own kingdom. Recent DNA studies suggest they're closer to animals than plants. Take note, vegetarians.

When we started hiring contractors to wrap our hay for silage — their machines juggling giant bales and mummifying them in plastic — we were asked if we wanted the hay inoculated. Many conventional farmers use inoculants to ensure the grasses break down into sweet silage, but our lucerne was breaking down very nicely on its own. Clearly we already had enough active fungi in the soil, no doubt because the 500 preparation was keeping the microbial life of the soil active.

Lately there seems to be a growing acknowledgement of the importance of fungi, with brochures and items in rural newspapers on fungi counts as an indicator of healthy ecosystems. But while there is a consensus that fungi are essential in nutrient cycles, and that without them whole ecosystems would die, we haven't really begun to understand their function and interaction.

- Crustose lichen
- Bracket fungi 
- Agaric fungus 
- Bolete fungus 
- Foliose lichen

It's said that naming is an imperial gesture, and that we need to move beyond it. I don't agree. I think we need to enrich our vocabulary even more. In particular, we need a new word for *bush*, which has become pejorative, an insult. *Just bush* is the dismissive term of every bad developer, as if they're doing us a favour with their bulldozers. As if in bushland there is never anything of value, anything to be learned, anything worth remembering.

Our relationship to land, including our ownership of it, ought not to be guided purely by profit and loss.

Solanum brownii
violet nightshade

Feral animals assist the seeds of feral plants by ploughing the ground for them. My vision of the farm's future does not include weeds, but I can't escape them. Some — blackberries, Noogoora burr, Bathurst burr and tiger pear, for instance — we regularly attack, chip, hoe, burn, but never manage to eliminate. Others have become part of the farm plan; they get used and even appreciated.

I've heard many stories of farmers spraying what they thought was silverleaf nightshade, a designated weed, only to discover it was the native plant *Solanum brownii*, violet nightshade, a pretty, purple-flowering, velvety shrub with tomato-like fruit. Travis says that with all his travelling through the valley, he's never seen silverleaf nightshade as a problem. In fact, he's hardly ever seen a silverleaf nightshade.

The dreary *Acacia paradoxa* is another native frequently mistaken for a weed. It's a prickly, short-lived, straggly bush that you wouldn't want in your garden. Even government weed inspectors have been known to poison it.

There are ducks on the top branch of this old ironbark. Do they nest up there? This could be why the goanna climbs the trunk every day. When I come around the bend it scuttles up the bark on the opposite side of the tree, but if I'm quiet it will stay where it prefers to be — on the sunny side. It has to be more than two metres long

Brush against stinging nettle and you soon know about it. What begins as a tingle becomes a rush of discomfort that lasts a long time, and you try to remain calm so you won't make it worse. That's probably why many farmers get nettled about nettles and attack them with herbicides.

But there's another side to the nettle story. I've developed an affection for *Urtica incisa*. I thought we had two varieties of stinging nettle growing on the farm, the English kind and a smaller local, but I was wrong. Our nettles are all dinky-di Australian, any difference in size a consequence of soil type, moisture and shade.

In the Middle Ages nettle was used as a fibre, but I use it as a fertiliser. One spring, overwhelmed by the proliferation of nettle around the cattle yards, we harvested it for a compost heap and it quickly became a mountain of worms. I've never had better compost for root vegetables — a testament to the presence of nitrogen.

Urtica incisa
stinging nettle

Guests are unenthusiastic at the thought of nettle soup, yet it holds its colour better than spinach and silverbeet, so I add it to spinach and fetta pies and to quiches. It's best to pick it before it gets too tall and woody or starts to flower. Pick it with your gloves on and don't tell your visitors they're eating it.

Feathery, fragrant fennel grows in the olive grove, along the road verges and beside the river. There are times when it's a two-metre crop, yet we haven't sown a single seed. I use all of it — the stems, shoots, flowers, pollen, bulb and dried seed. The herbalist Dorothy Hall, who grades herbs for their healing value, gives fennel five stars: it's good for the liver, gallbladder and pancreas. I never walk past fennel without nibbling some, and neither do the cattle. When they're moved to a fresh paddock they'll always munch the fennel first.

Foeniculum vulgare
fennel

I use the anise-flavoured seeds as a tea, I grind them up for curries, and mix them with cloves and add them to stir-fries. I once saw an old man in Umbria collecting fennel pollen to sprinkle on his lunch, and so now I rub it over fish cakes, yabbies and meat. Bees aren't particularly interested in fennel pollen but they suck the nectar. That's probably why I always enjoy the fragrance of February honey.

One of the prettiest vegetables growing in our pasture and olive grove is *Tragopogon porrifolius* — salsify, or oyster plant. Their grey-green shoots can be snapped off and eaten like spinach. You can steam them or toss them in olive oil, or use them as salad greens. A few weeks on, after rain, we harvest the roots. Aurora lifts them easily from the ground and brings bunches to the kitchen to be scrubbed and cut into chunks. I soak them in vinegar or lemon water so they won't turn grey, and steam or bake them. In a fancier mood I might make a béchamel sauce to pour over them.

Once salsify flowers, its roots become tough and less tasty. Leave them in the ground and they bloom into daisy-like flowers that tinge the hills a pink-purple at sunrise. Finally, before their life cycle ends, they form perfectly spherical seed heads, like large dandelion puffs.

Tragopogon porrifolius
salsify

The mammoth, variegated thistle that thrives in disturbed soil and new pasture is called *Silybum marianum*. Silly bum — why hasn't the name caught on? Like Scotch thistle, its flowers provide abundant nectar and pollen, and cattle chew the prickly flowers and stems as they dry off. Horses too seek them out. Silly bum — what a brand name that would make — is one of the group of plants used in liver tonics for humans.

Whenever we plough a paddock *Rapistrum rugosum*, or turnip weed, germinates triumphantly. If only intended crops were as lush. At its height, when in full flower, it can become impenetrable, but not to bees. You can hear their contagious excitement as you approach. Related to brassicas (like broccoli and mustard), turnip weed has little yellow flowers whose pollen is perfect for new bees preparing the hive for another season. Before it blossoms, cattle regard it as a delicacy.

Sisymbrium irio, commonly called London rocket, looks like a pale relative of turnip weed but tastes like arugula, and is good tossed with oil and lemon. It isn't a weed at all, but a herb which is native to many parts of the world, including Australia.

The Australian bush is a nocturnal world, and if you allow your eyes to adjust to the darkness you can see a great deal. The faint luminosity of the hills, the torchlike eyes of foxes. Elmswood comes alive in the light of the moon and stars. Everywhere spiders weave their remarkable fabrics, catching on your face and in your hair. Kangaroos congregate and nibble. Night birds swoop.

We make hay at night too, sometimes by moonlight.

Elmswood is constantly surprising me. This morning I opened a package to find bars of soap made from our olive oil and honey by a woman who runs a small soap factory in Brogo, near Bega, using solar power and rainwater.

So many people are involved in the work of this place. There's Gavin 'Scrubba' MacCallum and Simon Deery, the double act who manage the cattle with horse skills I'll never possess. Having arrived at sunrise with a motley assortment of working dogs, they're retrieving three hundred cows from the back boundary; they'll then sort these in the yards and will still be laughing at the end of the day. The olive pickers are due any minute; we've released a flock of sheep into the grove so that they can munch the grass before the pickers start. Dear Keith, now seventy-five, who used to help in the garden and the grove, has called in with some carp he's caught. Now I can try to perfect my carp pâté. And best of all, Aurora has just presented me with some strange but very significant biscuits . . .

Panicum effusum, hairy panic, is one of my favourite grasses. It's so delicate and wispy it responds to every breath of wind. Its toothpick stems snap and the seed heads cartwheel across the paddocks, eventually entangling in the fence wire. Some days, for as far as you can see, the fenceline is transformed, its barbed wire and wooden posts embroidered with gold thread.

On learning that Indigenous people used to harvest *Panicum*, Aurora disappears for an hour or so, returning with her arms full of grass. Donning headphones so she can listen to an audio book, she begins patiently to separate the seeds from the stems. As the seeds are the tiniest imaginable, it takes a long time to fill the coffee grinder. A few seconds' grinding produces a small amount of flour to which she inexplicably adds a little chilli. After a splash of milk, she moulds her mixture into biscuits and bakes them for a quarter of an hour.

While it's hard to be effusive about the taste of her *Panicum effusum* biscuits, I cannot begin to describe my pride in the cook. There can't have been many occasions in the past century when this plant has been used for human food, and Aurora's biscuits make me feel very hopeful.

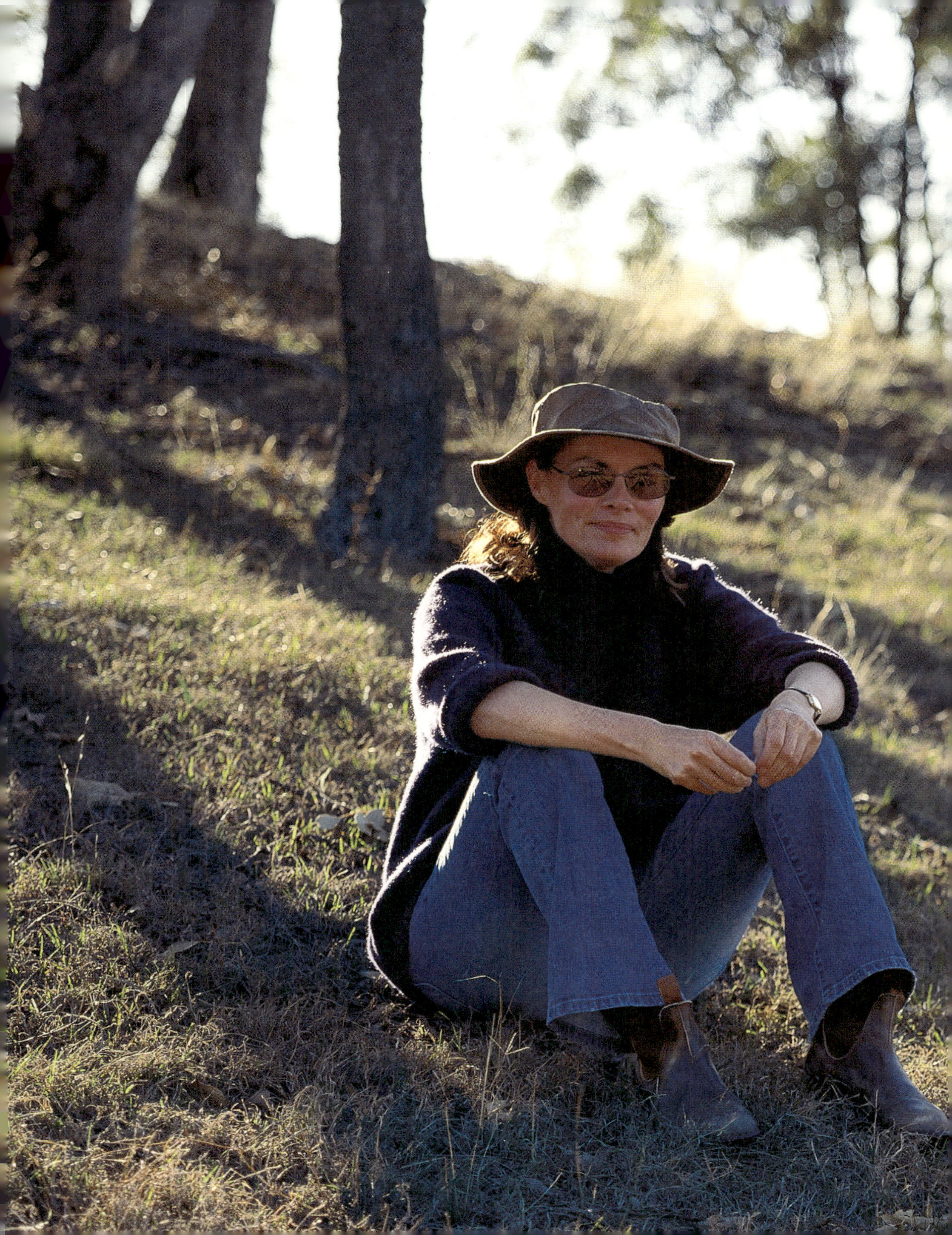

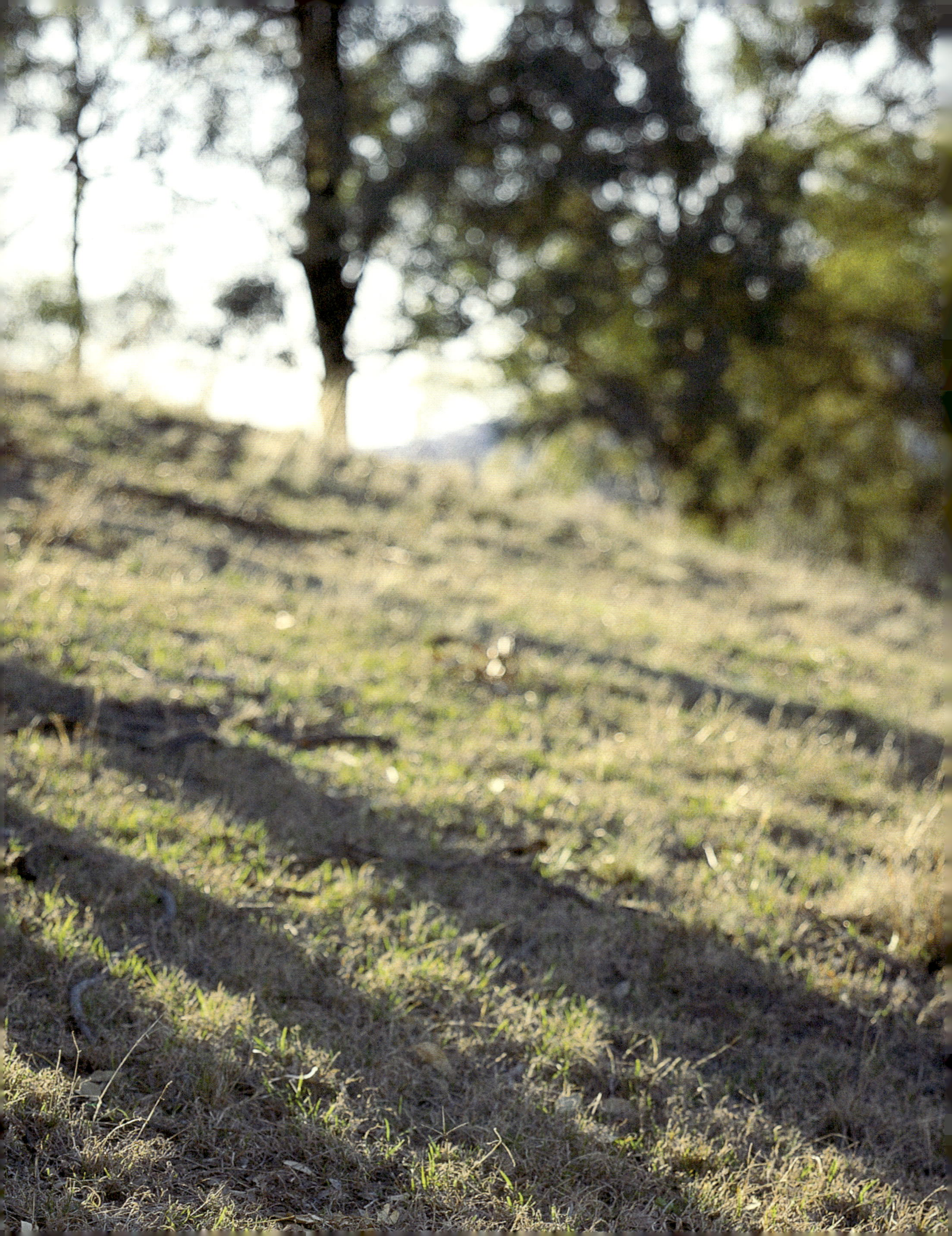

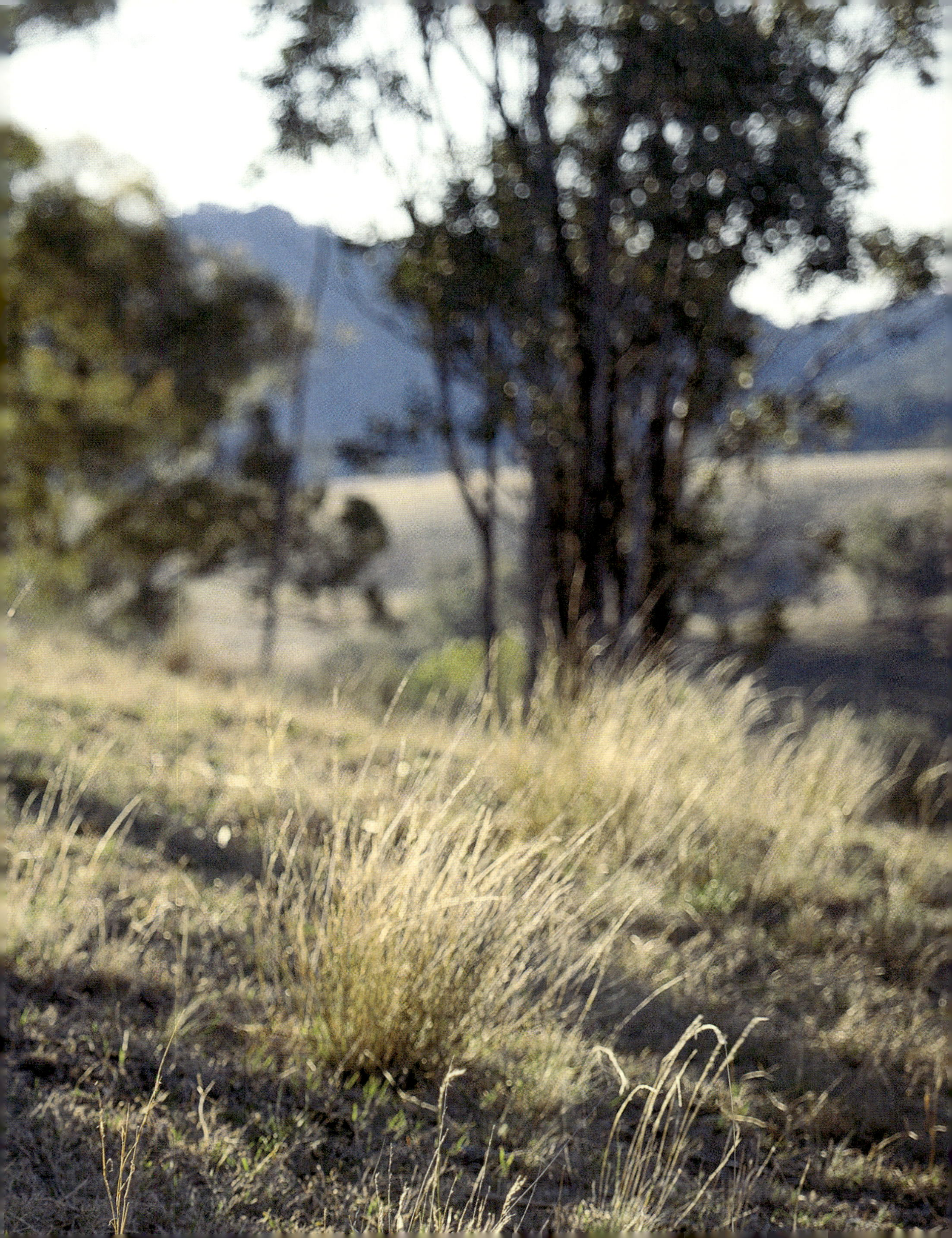

The Flora of Elmswood

	Family	Genus/species	Common name (NSW)
		(an asterisk denotes a plant not native to Elmswood; dark type indicates specimens found at Kewell Creek)	
FERNS Class: Filicopsida	ADIANTACEAE	*Adiantum aethiopicum*	common maidenhair
		Adiantum formosum	giant maidenhair
		Adiantum hispidulum	rough maidenhair
		Cheilanthes distans	**bristly cloak fern**
		Cheilanthes sieberi* subsp. *sieberi	**poison rock fern**
		Pellaea falcata	sickle fern
		Pellaea nana	sickle fern
		Pellaea paradoxa	
	ASPLENIACEAE	*Asplenium flabellifolium*	necklace fern
	AZOLLACEAE	*Azolla pinnata*	Pacific azolla
	BLECHNACEAE	*Doodia aspera*	prickly rasp fern
		Doodia australis	
		Doodia media	common rasp fern
	DENNSTAEDTIACEAE	*Pteridium esculentum*	bracken
	DICKSONIACEAE	*Calochlaena dubia*	common ground fern
	POLYPODIACEAE	*Pyrrosia rupestris*	rock felt fern
	PTERIDACEAE	*Pteris tremula*	tender brake
CYCADS Class: Cycadopsida	ZAMIACEAE	*Macrozamia concinna*	
CONIFERS Class: Coniferopsida	CUPRESSACEAE	*Callitris endlicheri*	black cypress pine
FLOWERING PLANTS Class: Magnoliopsida Subclass: Liliidae (Monocots)	ANTHERICACEAE	*Arthropodium milleflorum*	vanilla lily
		Arthropodium species B	vanilla lily
		Dichopogon fimbriatus	nodding chocolate lily
		Laxmannia gracilis	slender wire lily
		Tricoryne elatior	yellow autumn lily
	ASPHODELIACEAE	**Asphodelus fistulosus*	onion weed
	COLCHICACEAE	*Wurmbea biglandulosa*	early Nancy
	COMMELINACEAE	*Commelina cyanea*	scurvy weed
	CYPERACEAE	*Carex appressa*	tall sedge
		Carex inversa	knob sedge
		Carex longebrachiata	Bergalia tussock
		**Cyperus eragrostis*	umbrella sedge
		Cyperus gracilis	**sedge**
		Fimbristylis dichotoma	common fringe-rush

158
Ten Thousand Acres
A Love Story

Family	Genus/species	Common name (NSW)

(an asterisk denotes a plant not native to Elmswood; dark type indicates specimens found at Kewell Creek)

Family	Genus/species	Common name (NSW)
CYPERACEAE (cont.)	Gahnia aspera	rough saw-sedge
	Lepidosperma laterale	variable saw-sedge
IRIDACEAE	*Romulea rosea var. australis	onion grass
JUNCACEAE	Juncus usitatus	
JUNCAGINACEAE	Triglochin procerum	water ribbons
LEMNACEAE	Lemna sp.	duckweed
LOMANDRACEAE	Lomandra confertifolia	mat-rush
	Lomandra filiformis	wattle mat-rush
	Lomandra glauca	pale mat-rush
	Lomandra longifolia	spiny-headed mat-rush
	Lomandra multiflora subsp. multiflora	many-flowered mat-rush
LUZURIAGACEAE	Eustrephus latifolius	wombat berry
	Geitonoplesium cymosum	scrambling lily
ORCHIDACEAE	Acianthus fornicatus	pixie caps
	Caladenia carnea var. carnea	pink fairy
	Cymbidium canaliculatum	tiger orchid
	Dendrobium linguiforme	tongue orchid
	Dendrobium speciosum	rock lily
	Microtis unifolia	common onion orchid
	Pterostylis nutans	nodding greenhood
	Thelymitra ixioides	spotted sun orchid
PHORMIACEAE	Dianella caerulea var. caerulea	blue flax lily
	Dianella revoluta var. revoluta	blue flax lily
POACEAE	Aristida jerichoensis var. jerichoensis	Jericho wiregrass
	Aristida ramosa	**wiregrass**
	Austrodanthonia fulva	wallaby grass
	Austrostipa scabra	**corkscrew grass**
	Austrostipa verticillata	**slender bamboo grass**
	*Avena fatua	wild oats
	*Axonopus affinis	narrow-leaved carpet grass
	Bothriochloa macra	red grass
	*Briza maxima	quaking grass
	*Briza minor	shivery grass
	*Bromus cartharticus	prairie grass
	*Bromus diandrus	great brome
	*Chloris gayana	Rhodes grass
	Chloris truncata	**windmill grass**
	Chloris ventricosa	tall chloris
	Cymbopogon refractus	barbed wire grass
	Cynodon dactylon	couch
	Dichanthium sericeum subsp. sericeum	**Queensland bluegrass**
	Dichelachne micrantha	shorthair plumegrass

Pterostylis nutans
nodding greenhood

Thelymitra ixioides
spotted sun orchid

Family	Genus/species	Common name (NSW)
	(an asterisk denotes a plant not native to Elmswood; dark type indicates specimens found at Kewell Creek)	
POACEAE (cont.)	*Digitaria sanguinalis	summer grass
	Echinopogon caespitosus var. caespitosus	tufted hedgehog grass
	Echinopogon ovatus	forest hedgehog grass
	*Ehrharta erecta	panic veldtgrass
	*Eleusine tristachya	goose grass
	Elymus scaber var. scaber	**common wheatgrass**
	Eragrostis brownii	**Brown's lovegrass**
	Eragrostis leptostachya	paddock lovegrass
	Eriochloa pseudoacrotricha	**early spring grass**
	*Hordeum leporinum	barley grass
	Imperata cylindrica var. major	blady grass
	***Lolium perenne**	**perennial ryegrass**
	*Lolium rigidum	Wimmera ryegrass
	Microlaena stipoides var. stipoides	weeping grass
	Oplismenus aemulus	basket grass
	Oplismenus imbecillis	basket grass
	Panicum effusum	**hairy panic**
	Panicum queenslandicum var. queenslandicum	Yadbila grass
	Paspalidium distans	Warrego summergrass
	*Paspalum dilatatum	paspalum
	Pennisetum alopecuroides	swamp foxtail
	*Pennisetum clandestinum	kikuyu grass
	*Phalaris sp.	phalaris
	Phragmites australis	common reed
	Poa sieberiana var. sieberiana	snowgrass
	*Setaria gracilis	slender pigeon grass
	Sporobolus creber	**slender rat's tail grass**
	Themeda australis	kangaroo grass
SMILACACEAE	Smilax australis	sarsaparilla
TYPHACEAE	Typha orientalis	broad-leaved cumbungi
XANTHORRHOEACEAE	Xanthorrhoea glauca subsp. glauca	grass tree

Pennisetum alopecuroides swamp foxtail

FLOWERING PLANTS
Class: Magnoliopsida
Subclass: Magnoliidae
(Dicots)

Family	Genus/species	Common name (NSW)
ACANTHACEAE	Brunoniella australis	blue trumpet
	Rostellularia adscendens subsp. adscendens	red drums
AMARANTHACEAE	**Alternanthera pungens*	khaki weed
	Nyssanthes diffusa	barbwire weed
ANACARDIACEAE	*Schinus areira	pepper tree
APIACEAE	*Cyclospermum leptophyllum	slender celery
	Daucus glochidiatus	**native carrot**
	*Foeniculum vulgare	fennel
	Hydrocotyle laxiflora	stinking pennywort
APOCYNACEAE	Parsonsia eucalyptophylla	gargaloo
	Parsonsia lanceolata	northern silkpod
ASCLEPIADACEAE	*Araujia sericifera	moth plant
	*Gomphocarpus fruticosus	narrow-leaved cotton-bush
	Marsdenia rostrata	common milk vine
	Tylophora barbata	bearded tylophora
ASTERACEAE	*Artemisia verlotiorum	Chinese wormwood

Family	Genus/species	Common name (NSW)
	(an asterisk denotes a plant not native to Elmswood; dark type indicates specimens found at Kewell Creek)	
ASTERACEAE (cont.)	*Bidens pilosa*	cobbler's pegs
	Bidens subalternans	greater beggar's ticks
	Brachyscome sp.	
	Brachyscome multifida	**cut-leaved daisy**
	Calotis cuneifolia	purple burr-daisy
	Calotis lappulacea	**yellow burr-daisy**
	Carthamus lanatus	saffron thistle
	Cassinia arcuata	sifton bush
	Cassinia quinquefaria	cough-bush
	Centaurea calcitrapa	star thistle
	Chrysocephalum apiculatum	common everlasting
	Cirsium vulgare	spear thistle
	***Conyza sumatrensis**	**tall fleabane**
	Cymbonotus lawsonianus	bear's ear
	Facelis retusa	
	Glossogyne tannensis	cobbler's tack
	Hypochaeris radicata	catsear
	Lactuca serriola	prickly lettuce
	Minuria leptophylla	**minnie daisy**
	Olearia elliptica subsp. *elliptica*	sticky daisy bush
	Ozothamnus diosmifolius	white dogwood
	Pseuderanthemum variabile	**pastel flower**
	Rhodanthe anthemoides	chamomile sunray
	Senecio hispidulus var. *dissectus*	hill fireweed
	***Senecio madagascariensis**	**fireweed**
	Senecio quadridentatus	cotton fireweed
	Sigesbeckia australiensis	
	Sigesbeckia orientalis subsp. *orientalis*	Indian weed
	Silybum marianum	variegated thistle
	Solenogyne bellioides	solenogyne
	Soliva sessilis	jo-jo
	Sonchus oleraceus	common sowthistle
	Tagetes minuta	stinking Roger
	Taraxacum officinale	dandelion
	Tragopogon porrifolius	salsify
	Vernonia cinerea var. *cinerea*	
	Vittadinia sulcata	fuzzweed
	Xanthium occidentale	Noogoora burr
	Xanthium spinosum	Bathurst burr
BIGNONIACEAE	*Pandorea pandorana* subsp. *pandorana*	wonga wonga vine
BORAGINACEAE	*Cynoglossum australe*	Australian hound's tongue
	Echium plantagineum	Paterson's curse
BRASSICACEAE	*Capsella bursa-pastoris*	shepherd's purse
	Lepidium africanum	common peppercress
	Lepidium pseudohyssopifolium	peppercress
	Rapistrum rugosum	turnip weed
	Rorippa nasturtium-aquaticum	watercress
	Sisymbrium irio	London rocket
CACTACEAE	*Opuntia aurantiaca*	tiger pear
	Opuntia stricta var. *stricta*	common prickly pear

Family	Genus/species	Common name (NSW)
	(an asterisk denotes a plant not native to Elmswood; dark type indicates specimens found at Kewell Creek)	
CAMPANULACEAE	Wahlenbergia communis	tufted bluebell
	Wahlenbergia gracilis	sprawling bluebell
	Wahlenbergia sp.	**native bluebell**
	Wahlenbergia stricta	tall bluebell
CARYOPHYLLACEAE	*Paronychia brasiliana	Chilean whitlow wort
	*Petrorhagia nanteuilii	proliferous pink
	***Petrorhagia velutina**	**velvet pink**
	*Silene gallica var. gallica	campion
	*Stellaria media	common chickweed
	Stellaria pungens	prickly starwort
CASUARINACEAE	Allocasuarina torulosa	forest oak
	Casuarina cunninghamiana subsp. cunninghamiana	river she-oak
CHENOPODIACEAE	*Chenopodium album	fat hen
	Einadia hastata	berry saltbush
	Einadia nutans	climbing saltbush
	Einadia trigonos	**fishweed**
	Maireana microphylla	eastern cottonbush
CLUSIACEAE	Hypericum gramineum	small St John's wort
	*Hypericum perforatum	St John's wort
CONVOLVULACEAE	Convolvulus erubescens	bindweed
	Dichondra repens	**kidney weed**
CRASSULACEAE	Crassula sieberiana	Australian stonecrop
CUNONIACEAE	Aphanopetalum resinosum	gum vine
DILLENIACEAE	Hibbertia obtusifolia	Guinea flower
	Hibbertia scandens	climbing Guinea flower
DROSERACEAE	Drosera auriculata	sundew
EPACRIDACEAE	Lissanthe strigosa	peach heath
	Melichrus urceolatus	urn heath
EUPHORBIACEAE	Breynia oblongifolia	coffee bush
	Chamaesyce drummondii	caustic weed
	Phyllanthus hirtellus	thyme spurge
	Phyllanthus occidentalis	spurge
	Phyllanthus virgatus	**spurge**
FABACEAE-FABOIDEAE	Daviesia genistifolia	broom bitter pea
	Desmodium brachypodum	large tick-trefoil
	Desmodium varians	**slender tick-trefoil**
	Glycine clandestina	twining glycine
	Glycine tabacina	**glycine**
	Hardenbergia violacea	false sarsaparilla
	Hovea lanceolata	lance-leaf hovea
	Indigofera australis	native indigo
	*Medicago polymorpha	burr medic
	*Medicago sativa	lucerne
	*Melilotus indicus	Hexham scent
	Swainsona galegifolia	smooth darling pea
	*Trifolium arvense	haresfoot clover
	*Trifolium campestre	hop clover
	*Trifolium dubium	yellow suckling clover
	*Trifolium pratense	red clover

Family	Genus/species	Common name (NSW)
	(an asterisk denotes a plant not native to Elmswood; dark type indicates specimens found at Kewell Creek)	
FABACEAE-FABOIDEAE (cont.)	*Trifolium repens*	white clover
	Trifolium subterraneum	subterranean clover
	Trigonella suavissima	Cooper's clover
	Vicia sativa	vetch
FABACEAE-MIMOSOIDEAE	Acacia brownii	prickly Moses
	Acacia crassa subsp. crassa	curracabah
	Acacia decora	western golden wattle
	Acacia falcata	hickory wattle
	Acacia farnesiana	mimosa bush
	Acacia implexa	hickory wattle
	Acacia maidenii	Maiden's wattle
	Acacia paradoxa	kangaroo thorn
	Acacia salicina	cooba
FUMARIACEAE	*Fumaria muralis subsp. muralis*	wall fumitory
GERANIACEAE	*Geranium molle*	cranesbill geranium
	Geranium solanderi var. solanderi	**native geranium**
	Pelargonium inodorum	**wild geranium**
GOODENIACEAE	Goodenia hederacea subsp. hederacea	ivy goodenia
	Goodenia paniculata	swamp goodenia
HALORAGACEAE	Gonocarpus elatus	hill raspwort
LAMIACEAE	Ajuga australis	austral bugle
	Lamium amplexicaule	dead nettle
	Marrubium vulgare	horehound
	Mentha satureioides	**creeping mint**
	Plectranthus parviflorus	cockspur flower
	Stachys arvensis	stagger weed
LAURACEAE	Cassytha pubescens	devil's twine
LOBELIACEAE	Pratia purpurascens	whiteroot
LORANTHACEAE	Amyema cambagei	she-oak mistletoe
	Amyema miquelii	box mistletoe
	Amyema pendulum subsp. pendulum	drooping mistletoe
	Dendrophthoe vitellina	
MALVACEAE	Abutilon oxycarpum	flannel weed
	Hibiscus sturtii var. sturtii	hill hibiscus
	Sida cunninghamii	
	Sida filiformis	
	Sida rhombifolia	Paddy's lucerne
MELIACEAE	Melia azedarach	white cedar
	Toona ciliata	red cedar
MENISPERMACEAE	Stephania japonica var. discolor	snake vine
MORACEAE	Ficus coronata	creek sandpaper fig
	Ficus rubiginosa f. rubiginosa	rusty fig
MYOPORACEAE	Eremophila debilis	winter apple
	Myoporum montanum	western boobialla
MYRSINACEAE	Rapanea variabilis	muttonwood
MYRTACEAE	Angophora floribunda	rough-barked apple
	Callistemon sieberi	river bottlebrush
	Eucalyptus albens	white box
	Eucalyptus blakelyi	Blakely's red gum

Acacia decora
western golden wattle

Sida cunninghamii

Family	Genus/species	Common name (NSW)
	(an asterisk denotes a plant not native to Elmswood; dark type indicates specimens found at Kewell Creek)	
MYRTACEAE (cont.)	*Eucalyptus bridgesiana*	apple box
	Eucalyptus crebra	narrow-leaved ironbark
	Eucalyptus goniocalyx	bundy
	Eucalyptus laevopinea	silver-top stringybark
	Eucalyptus melliodora	yellow box
	Eucalyptus nobilis	ribbon gum
	Eucalyptus punctata	grey gum
	Eucalyptus tereticornis	forest red gum
	Leptospermum polygalifolium	yellow tea tree
NYCTAGINACEAE	*Boerhavia dominii*	tarvine
OLEACEAE	*Notelaea microcarpa* var. *microcarpa*	native olive
ONAGRACEAE	**Oenothera indecora* subsp. *bonariensis*	evening primrose
	**Oenothera stricta* subsp. *stricta*	evening primrose
OXALIDACEAE	**Oxalis exilis**	**wood sorrel**
PAPAVERACEAE	**Argemone ochroleuca* subsp. *ochroleuca*	Mexican poppy
	**Eschscholzia californica*	California poppy
	**Papaver somniferum* subsp. *setigerum*	opium poppy
PHYTOLACCACEAE	**Phytolacca octandra*	inkweed
PITTOSPORACEAE	*Bursaria spinosa* subsp. *spinosa*	blackthorn
	Hymenosporum flavum	native frangipani
	Pittosporum multiflorum	orange thorn
	Pittosporum revolutum	hairy pittosporum
	Pittosporum undulatum	sweet pittosporum
PLANTAGINACEAE	*Plantago debilis*	slender plantain
	**Plantago lanceolata*	lamb's tongues
POLYGONACEAE	*Persicaria decipiens*	slender knotweed
	Rumex brownii	**swamp dock**
	**Rumex crispus*	curled dock
	**Rumex obtusifolius* subsp. *obtusifolius*	broadleaf dock
PORTULACACEAE	***Portulaca oleracea***	**pigweed**
PRIMULACEAE	****Anagallis arvensis***	**scarlet pimpernel**
PROTEACEAE	*Persoonia linearis*	narrow-leaved geebung
RANUNCULACEAE	*Clematis aristata*	old man's beard
	Clematis glycinoides var. *glycinoides*	headache vine
	Ranunculus lappaceus	common buttercup
RHAMNACEAE	*Cryptandra amara* var. *amara*	cryptandra
ROSACEAE	*Acaena novae-zelandiae*	bidgee-widgee
	**Rosa rubiginosa*	sweet briar
	**Rubus fruticosus* sp. agg.	blackberry
	Rubus moluccanus var. *trilobus*	Molucca bramble
	Rubus parvifolius	native raspberry
	**Sanguisorba minor* subsp. *muricata*	sheep's burnet
RUBIACEAE	*Asperula conferta*	common woodruff
	Canthium odoratum	shiny-leaved canthium
	Galium propinquum	Maori bedstraw
	Morinda jasminoides	morinda

Bursaria spinosa subsp. *spinosa*
blackthorn

Family	Genus/species	Common name (NSW)
	(an asterisk denotes a plant not native to Elmswood; dark type indicates specimens found at Kewell Creek)	
RUTACEAE	*Richardia stellaris*	field madder
	Correa reflexa var. reflexa	common correa
	Geijera parviflora	wilga
SANTALACEAE	Choretrum species A	sour bush
	Exocarpos cupressiformis	native cherry
	Santalum lanceolatum	northern sandalwood
SAPINDACEAE	Alectryon oleifolius subsp. elongatus	western rosewood
	Dodonaea viscosa subsp. cuneata	wedge-leaf hop bush
	Dodonaea viscosa subsp. mucronata	hop bush
	Dodonaea viscosa subsp. spatulata	hop bush
SCROPHULARIACEAE	*Linaria pelisseriana	Pelisser's toadflax
	*Verbascum thapsus subsp. thapsus	blanket weed
	*Verbascum virgatum	twiggy mullein
	Veronica plebeia	trailing speedwell
SOLANACEAE	*Lycium ferocissimum	African boxthorn
	Nicotiana suaveolens	native tobacco
	Solanum brownii	violet nightshade
	Solanum campanulatum	nightshade
	Solanum cinereum	Narrawa burr
	Solanum opacum	green-berry nightshade
	Solanum prinophyllum	forest nightshade
STACKHOUSIACEAE	Stackhousia viminea	slender stackhousia
STERCULIACEAE	Brachychiton populneus subsp. populneus	kurrajong
THYMELAEACEAE	Pimelea latifolia subsp. elliptifolia	rice flower
	Pimelea linifolia	**slender rice flower**
ULMACEAE	Celtis australis	nettle tree
	Trema tomentosa var. viridis	native peach
URTICACEAE	Urtica incisa	stinging nettle
VERBENACEAE	Clerodendrum tomentosum	hairy clerodendrum
	*Verbena bonariensis	purpletop
	***Verbena litoralis**	**coastal verbena**
	*Verbena rigida var. rigida	veined verbena
VIOLACEAE	Hymenanthera dentata	tree violet
	Viola betonicifolia	showy violet
	Viola hederacea	ivy-leaved violet
VISCACEAE	Notothixos cornifolius	kurrajong mistletoe
VITACEAE	Cayratia clematidea	slender grape
	Cissus antarctica	water vine
	Cissus opaca	small-leaved water vine

Brachychiton populneus subsp. *populneus*
kurrajong

This list includes a large number of non-native species that occur exclusively along six kilometres of the Pages River

The Birds of Elmswood

Family	Genus/species	Common name (NSW)
	(an asterisk denotes a bird not native to Elmswood)	

Class: Aves

Family	Genus/species	Common name (NSW)
PHASIANIDAE	*Coturnix pectoralis*	stubble quail
ANATIDAE	*Chenonetta jubata*	Australian wood duck
	Anas superciliosa	Pacific black duck
	Anas gracilis	grey teal
PELECANIDAE	*Pelecanus conspicillatus*	Australian pelican
ARDEIDAE	*Egretta novaehollandiae*	white-faced heron
	Ardea ibis	cattle egret
THRESKIORNITHIDAE	*Threskiornis spinicollis*	straw-necked ibis
	Platelea regia	royal spoonbill
ACCIPITRIDAE	*Elanus notatus*	black-shouldered kite
	Haliastur sphenurus	whistling kite
	Haliaeetus leucogaster	white-bellied sea-eagle
	Accipiter fasciatus	brown goshawk
	Aquila audax	wedge-tailed eagle
FALCONIDAE	*Falco peregrinus*	peregrine falcon
	Falco cenchroides	nankeen kestrel
RALLIDAE	*Fulica atra*	Eurasian coot
RECURVIROSTRIDAE	*Himantopus himantopus*	black-winged stilt
CHARADRIIDAE	*Vanellus miles*	masked lapwing
COLUMBIDAE	*Phaps chalcoptera*	common bronzewing
	Ocyphaps lophotes	crested pigeon
	Geopelia striata	peaceful dove
	Geopelia humeralis	bar-shouldered dove
CACATUIDAE	*Cacatua roseicapilla*	galah
	Cacatua tenuirostris	long-billed corella
	Cacatua galerita	sulphur-crested cockatoo
PSITTACIDAE	*Alisterus scapularis*	Australian king-parrot
	Platycercus elegans	crimson rosella
	Platycercus eximius	eastern rosella
	Psephotus haematonotus	red-rumped parrot
CUCULIDAE	*Cacomantis flabelliformis*	fan-tailed cuckoo
	Eudynamis scolopacea	common koel
	Scythrops novaehollandiae	channel-billed cuckoo
CENTROPODIDAE	*Centropus phasianinus*	pheasant coucal
STRIGIDAE	*Ninox novaeseelandiae*	southern boobook
TYTONIDAE	*Tyto novaehollandiae*	masked owl
	Tyto alba	barn owl
PODARGIDAE	*Podargus strigoides*	tawny frogmouth

Family	Genus/species	Common name (NSW)
	(an asterisk denotes a bird not native to Elmswood)	
CAPRIMULGIDAE	*Eurostopodus mystacalis*	white-throated nightjar
HALCYONIDAE	*Dacelo novaeguineae*	laughing kookaburra
	Todiramphus sanctus	sacred kingfisher
MEROPIDAE	*Merops ornatus*	rainbow bee-eater
CORACIIDAE	*Eurystomus orientalis*	dollarbird
CLIMACTERIDAE	*Corombates leucophaeus*	white-throated treecreeper
MALURIDAE	*Malurus cyaneus*	superb fairy-wren
PARDALOTIDAE	*Pardalotus punctatus*	spotted pardalote
	Pardalotus striatus	striated pardalote
ACANTHIZIDAE	*Gerygone olivacea*	white-throated gerygone
	Acanthiza pusilla	brown thornbill
MELIPHAGIDAE	*Anthochaera carunculata*	red wattlebird
	Philemon corniculatus	noisy friarbird
	Entomyzon cyanotis	blue-faced honeyeater
	Manorina melanocephala	noisy miner
	Meliphaga lewinii	Lewin's honeyeater
	Lichenostomus chrysops	yellow-faced honeyeater
	Acanthorhynchus tenuirostris	eastern spinebill
PETROICIDAE	*Microeca leucophaea*	jacky winter
	Petroica rosea	rose robin
	Eopsaltria australis	eastern yellow robin
CINCLOSOMATIDAE	*Psophodes olivaceus*	eastern whipbird
NEOSITTIDAE	*Daphoenositta chrysoptera*	varied sittella
PACHYCEPHALIDAE	*Pachycephala pectoralis*	golden whistler
	Pachycephala rufiventris	rufous whistler
	Colluricincla harmonica	grey shrike-thrush
DICRURIDAE	*Myiagra cyanoleuca*	satin flycatcher
	Grallina cyanoleuca	peewee
	Rhipidura fuliginosa	grey fantail
	Rhipidura leucophrys	willie wagtail
	Dicrurus hottentottus	spangled drongo
CAMPEPHAGIDAE	*Coracina novaehollandiae*	black-faced cuckoo-shrike
ORIOLIDAE	*Oriolus sagittatus*	olive-backed oriole
ARTAMIDAE	*Cracticus torquatus*	grey butcherbird
	Cracticus nigrogularis	pied butcherbird
	Gymnorhina tibicen	Australian magpie
	Strepera graculina	pied currawong
CORVIDAE	*Corvus coronoides*	Australian raven
CORCORACIDAE	*Corcorax melanorhamphos*	white-winged chough
PTILONORHYNCHIDAE	*Ptilonorhynchus violaceus*	satin bowerbird
MOTACILIDAE	*Anthus novaeseelandiae*	Richard's pipit
PASSERIDAE	*Taeniopygia bichenovii*	double-barred finch
	Neochmia temporalis	red-browed finch
DICAEIDAE	*Dicaeum hirundinaceum*	mistletoebird
HIRUNDINIDAE	*Hirundo neoxena*	welcome swallow
	Hirundo ariel	fairy martin
ZOSTEROPIDAE	*Zosterops lateralis*	silvereye
STURNIDAE	**Sturnus vulgaris*	common starling
	**Acridotheres tristis*	common myna

Acknowledgements This book was made possible by Julie Gibbs, who said yes to a vague idea. Travis Peake helped me understand the botany and ecosystem of the farm and captured it with his photographs. Simon Griffiths' extraordinary eye also provided beautiful images. Meredith Rose's sensitive editing and Sandy Cull's design have been indispensable. Thanks also to John Gollon, Chris Reid, Daryl Cluff, David Marks and Bob Gulliford.

Writing always takes me away from the 'doing' of the farm. Thank you, Gavin Prescott, for keeping things running, Gavin MacCallum and Simon Deery for taking care of the stock, Yvonne Mitchell for keeping the office functioning, and Colin Watts for tending the garden. My daughter Aurora helps in so many ways, and if it weren't for Phillip I wouldn't be here in the first place.

Further information on native grasses can be found on the website of Stipa, the Australian native grasses association, www.stipa.com.au. For an inspiring practical book on how to farm without destroying native grasses, see *Farming without Farming* by Daryl Cluff (available from Stipa). Information about organics and biodynamics can be obtained from the Organics Directory, www.theorganicsdirectory.com.au. And to help bring all the ideas together, go to www.holisticresults.com.au.

LEFT TO RIGHT Gavin Prescott, Simon Deery, me, Gavin MacCallum, Colin Watts, Yvonne Mitchell, Phillip, Aurora

Photograph sources SIMON GRIFFITHS: pages ii–iii, 2–4, 7, 11, 12, 13 (bottom L), 15–16, 18–20, 22–5, 27, 29, 32–3, 35, 38–9, 43, 45, 48–50, 53–5, 57, 59–60, 64–6, 68–9, 71, 73–4, 78 (R), 81–2, 84–97, 99 (bottom L & R), 103–105, 108–12, 115, 116–17, 120–3, 127, 130, 132, 134–5, 138–42, 149–50, 154–7

TRAVIS PEAKE: pages vi–viii, 13 (top L & R, bottom R), 17, 21, 28, 30, 41, 44, 46, 56, 61–3, 70, 75, 77, 78 (L), 99 (top L & R), 100–101, 107, 119, 124–5, 128–9, 131, 133, 137, 144–8, 152–3, 159–60, 163–5

Albrecht Dürer's *The Great Piece of Turf* is reproduced on page 79 by permission of Graphische Sammlung Albertina, Vienna, Austria/Bridgeman Art Library. The photographs on pages 5 & 168 are by Adele Orton.